T0297468

Reconstruction and Analysis of 3D Scenes

Martin Weinmann

Reconstruction and Analysis of 3D Scenes

From Irregularly Distributed 3D Points
to Object Classes

 Springer

Martin Weinmann
Institute of Photogrammetry and Remote
 Sensing
Karlsruhe Institute of Technology
Karlsruhe
Germany

ISBN 978-3-319-29244-1   ISBN 978-3-319-29246-5   (eBook)
DOI 10.1007/978-3-319-29246-5

Library of Congress Control Number: 2016930271

Printed on acid-free paper

This Springer imprint is published by SpringerNature
The registered company is Springer International Publishing AG Switzerland

Reconstruction and Analysis of 3D Scenes

– From Irregularly Distributed 3D Points to Object Classes –

Zur Erlangung des akademischen Grades eines

Doktor-Ingenieurs (Dr.-Ing.)

von der

Fakultät für Bauingenieur-, Geo- und Umweltwissenschaften

des

Karlsruher Instituts für Technologie (KIT)

genehmigte
Dissertation

von

Dipl.-Ing. Martin Weinmann

aus Karlsruhe

Tag der mündlichen Prüfung: 24.07.2015

Referent: Prof. Dr.-Ing. habil. Stefan Hinz
Korreferent: Prof. Dr.-Ing. Uwe Stilla

Karlsruhe 2015

To my family

Foreword

Over the last decades, 3D point clouds have become a standard data source for automated navigation, mapping and reconstruction tasks. While, in the past, the acquisition of such data was restricted to highly specialized photogrammetric cameras or laser scanners, this situation changed drastically during the last few years. Many common off-the-shelf sensors of modern photogrammetry, remote sensing and computer vision provide 3D point cloud data meanwhile. The available sensor suite includes, for instance, stereo- or multiview vision sensors, laser scanners, time-of-flight cameras and structured light sensors.

Although raw point cloud data delivered by these sensors may give a visually striking impression of the recorded scene, the 3D points are usually unstructured, irregularly distributed and by far not inhering any semantics of the objects. This is the starting point of this work. It develops a coherent concept of 3D point cloud processing with the ultimate goal of identifying the objects present in the scene, given only a few training samples of the sought-after object classes. The modules of the developed system comprise highly sophisticated data processing methods, such as co-registration of different point clouds, geometric pattern analysis including optimal neighborhood selection, feature relevance estimation and feature selection as well as contextual classification. In addition, also the integration of imaging sensors, in particular thermal cameras, in the acquisition system is touched in this work. All methods are fully automated and almost free of parameters set by the user, except those defining the objects to classify.

The implemented concept is not restricted to a specific kind of object class like many other approaches are. Furthermore, it can be applied to large real-world scenes implying millions and millions of 3D points, thereby producing reasonable and high quality results—especially in the light of not needing any user interaction. Both from a scientific and from an application-oriented view, this work is a milestone and a big step forward towards the long-term goal of a fully automated mapping system applicable to complex outdoor scenes. Last but not least, a thorough evaluation based on typical real-world examples elaborates open issues to be tackled in future work.

Karlsruhe   Stefan Hinz
December 2015

Preface

The fully automatic processing and analysis of 3D point clouds represents a topic of major interest in the fields of photogrammetry, remote sensing, computer vision, and robotics. Focusing on general applicability and considering a typical end-to-end processing workflow from raw 3D point cloud data to semantic objects in the scene, we introduce a novel and fully automated framework involving a variety of components which cover (i) the filtering of noisy data, (ii) the extraction of appropriate features, (iii) the adequate alignment of several 3D point clouds in a common coordinate frame, (iv) the enrichment of 3D point cloud data with other types of information, and (v) the semantic interpretation of 3D point clouds. For each of these components, we reflect the fundamentals and related work, and we further provide a comprehensive description of our novel approaches with significant advantages. Based on a detailed evaluation of our framework on various benchmark datasets, we provide objective results for different steps in the processing workflow and demonstrate the performance of the proposed approaches in comparison to state-of-the-art techniques.

Target Audience

Although this book has been written as research monograph, the target audience includes a broad community of people who are dealing with 3D point cloud processing, reaching from students at undergraduate or graduate level to lecturers, practitioners and researchers in photogrammetry, remote sensing, computer vision and robotics. Background knowledge on 3D point cloud processing will certainly improve the reader's appreciation but is not mandatory since reviews on fundamentals and related work are provided for the different components of an end-to-end processing workflow from raw 3D point cloud data to semantic objects in the scene.

Difficulty

Beginner to intermediate and expert. This book is intended to address a broad community by reflecting the fundamentals and related work on the different components of an end-to-end processing workflow from raw 3D point cloud data to semantic objects in the scene. Additionally, a comprehensive description of novel approaches with significant advantages is provided, whereby a detailed evaluation of the proposed approaches in comparison to state-of-the-art techniques is carried out on various benchmark datasets. In this regard, beginners may obtain an impression about the performance of different approaches with respect to different evaluation criteria, whereas experts in the related fields may appreciate the novel approaches and be inspired to conduct further investigations.

Organization of the Content

This book addresses the reconstruction and analysis of 3D scenes and thereby focuses on the presentation of an end-to-end processing workflow from raw 3D point cloud data to semantic objects in the scene. To a reasonable extent, all the material about one topic is presented together in a separate chapter, making the book suitable as a reference work as well as a tutorial on different subtopics.

In Chap. 1, we provide an introduction which focuses on explaining the goals of this work, the main challenges when addressing these goals and the scientific contributions developed in this context. We additionally provide a list of our publications, take a glance view on our novel framework for advanced 3D point cloud processing from raw data to semantic objects and finally give an overview on the topics presented in the whole book.

In Chap. 2, we consider fundamentals in the form of general definitions that will be important for a consideration of our work in the global context. Since, in this work, we focus on the use of 3D point cloud data as input for our methodology, we briefly explain how such 3D point cloud data may be acquired indirectly from 2D imagery via passive techniques and how it may be acquired directly by involving active techniques. Particularly the active techniques offer many advantages and we therefore consider respective systems for data acquisition in the scope of this book. An analysis of such acquisition systems reveals that the acquired 3D point cloud data may easily be represented in the form of range and intensity images which, in turn, might facilitate a variety of applications in terms of simplicity and efficiency. In this regard, we consider point quality assessment where the aim is to quantify the quality of the single range measurements captured with the involved acquisition system, and we present and discuss two novel approaches and their potential in comparison to related work.

In Chap. 3, we briefly explain fundamental concepts for extracting features from 2D imagery and 3D point cloud data. In this regard, we derive a definition for the general term of a *feature* and consider different types of features that might occur in 2D imagery or 3D point cloud data. These fundamentals have become the basis and core of a plethora of applications in photogrammetry, remote sensing, computer vision and robotics, and they also provide the basis for the subsequent chapters.

In Chap. 4, we focus on transforming 3D point clouds—where each 3D point cloud has been acquired with respect to the local coordinate frame of the acquisition system—into a common coordinate frame. We consider related work and thereby particularly focus on efficient approaches relying on the use of keypoints. Based on these considerations, we present a novel framework for point cloud registration which exploits the 2D representations of acquired 3D point cloud data in the form of range and intensity images in order to quantify the quality of respective range measurements and subsequently reliably extract corresponding information between different scans. In this regard, we may consider corresponding information in the respective image representations or in 3D space, and we may also introduce a weighting of derived correspondences with respect to different criteria. Thus, the correspondences may serve as input for the registration procedure for which we present two novel approaches. An evaluation on a benchmark dataset clearly reveals the performance of our framework, and we discuss the derived results with respect to different aspects including registration accuracy and computational efficiency.

In Chap. 5, we focus on the enrichment of acquired 3D point clouds by additional information acquired with a thermal camera. We reflect related work and present a novel framework for thermal 3D mapping. Our framework involves a radiometric correction, a geometric calibration, feature extraction and matching, and two approaches for the co-registration of 3D and 2D data. For the example of an indoor scene, we demonstrate the performance of our framework in terms of both accuracy and applicability, and we discuss the derived results with respect to potential use-cases.

In Chap. 6, we focus on 3D scene analysis where the aim is to uniquely assign each 3D point of a given 3D point cloud a respective (semantic) class label. We provide a survey on related work and present a novel framework addressing neighborhood selection, feature extraction, feature selection and classification. Furthermore, we consider extensions of our framework toward large-scale capability and toward the use of contextual information. In order to demonstrate the performance of our framework, we provide an extensive experimental evaluation on publicly available benchmark datasets and discuss the derived results with respect to classification accuracy, computational efficiency and applicability of involved methods.

In Chap. 7, we summarize the contents of the book and provide concluding remarks as well as suggestions for future work.

Abstract

The fully automatic processing and analysis of 3D point clouds represents a topic of major interest in the fields of photogrammetry, remote sensing, computer vision, and robotics. Focusing on general applicability and considering a typical end-to-end processing workflow from raw 3D point cloud data to semantic objects in the scene, we introduce a novel and fully automated framework involving a variety of components which cover (i) the filtering of noisy data, (ii) the extraction of appropriate features, (iii) the adequate alignment of several 3D point clouds in a common coordinate frame, (iv) the enrichment of 3D point cloud data with other types of information, and (v) the semantic interpretation of 3D point clouds. For each of these components, we reflect the fundamentals and related work, and we further provide a comprehensive description of our novel approaches with significant advantages. Based on a detailed evaluation of our framework on various benchmark datasets, we provide objective results for different steps in the processing workflow and demonstrate the performance of the proposed approaches in comparison to state-of-the-art techniques.

In particular, our derived results reveal that (i) our presented point quality measures allow an appropriate filtering of noisy data and have a positive impact on the automated alignment of several 3D point clouds in a common coordinate frame, (ii) the extraction of appropriate features improves the automated alignment of several 3D point clouds with respect to accuracy and efficiency, and even allows a co-registration of 3D and 2D data acquired with different sensor types, (iii) our presented strategies for keypoint-based point cloud registration in terms of either projective scan matching or omnidirectional scan matching allow a highly accurate alignment of several 3D point clouds in a common coordinate frame, (iv) our presented strategies in terms of a RANSAC-based homography estimation and a projective scan matching allow an appropriate co-registration of 3D and 2D data acquired with different sensor types, and (v) our strategy for increasing the distinctiveness of low-level geometric features via the consideration of an optimal neighborhood for each individual 3D point and our strategy for only selecting compact and robust subsets of relevant and informative features have a significantly beneficial impact on the results of 3D scene analysis. Thus, our novel approaches allow an efficient reconstruction and analysis of large 3D environments up to city scale, and they offer a great potential for future research.

Kurzfassung

Die automatische Verarbeitung und Analyse von 3D-Punktwolken stellt in den Bereichen der Photogrammetrie, Fernerkundung, Computer Vision und Robotik ein wichtiges Thema dar. Im Hinblick auf eine allgemeine Anwendbarkeit wird in der vorliegenden Arbeit eine neue und vollautomatisierte Methodik vorgestellt, welche die wesentlichen Schritte von der Erfassung von 3D-Punktwolken bis hin zur Ableitung von semantischen Objekten in der Szene betrachtet. Diese Methodik umfasst verschiedene Komponenten, welche (i) die Filterung von verrauschten Daten, (ii) die Extraktion von geeigneten Merkmalen, (iii) die angemessene Ausrichtung von mehreren einzelnen 3D-Punktwolken in einem gemeinsamen Koordinatensystem, (iv) die Anreicherung von 3D-Punktwolken mit zusätzlicher Information und (v) die semantische Interpretation von 3D-Punktwolken umfassen. Für jede Komponente werden die Grundlagen sowie der aktuelle Stand der Forschung aufgezeigt. Ferner werden die im Rahmen dieser Arbeit entwickelten Verfahren mit deutlichen Vorteilen gegenüber den bisherigen Verfahren genauer beleuchtet. Basierend auf einer umfassenden Auswertung auf verschiedenen Standard-Datensätzen werden objektive Ergebnisse für verschiedene Schritte in der Datenverarbeitung präsentiert und die Leistungsfähigkeit der entwickelten Methodik im Vergleich zu Standard-Verfahren verdeutlicht.

Im Speziellen zeigen die im Rahmen der vorliegenden Arbeit erzielten Ergebnisse, (i) dass die entwickelten Qualitätsmaße eine angemessene Filterung von verrauschten Daten ermöglichen und sich positiv auf die automatische Ausrichtung von mehreren einzelnen 3D-Punktwolken in einem gemeinsamen Koordinatensystem auswirken, (ii) dass die Extraktion von geeigneten Merkmalen bei der automatischen Ausrichtung von mehreren 3D-Punktwolken sowohl die Genauigkeit als auch die Effizienz der getesteten Verfahren verbessert und sogar eine Ko-Registrierung von 3D- und 2D-Daten, welche mit verschiedenen Sensortypen erfasst wurden, ermöglicht, (iii) dass die vorgestellten Strategien zur Punkt-basierten Registrierung von 3D-Punktwolken über ein projektives Scan Matching und ein omnidirektionales Scan Matching zu einer sehr genauen automatischen Ausrichtung von einzelnen 3D-Punktwolken in einem gemeinsamen Koordinatensystem führen, (iv) dass die vorgestellten Strategien einer RANSAC-basierten Homographie-Schätzung und eines projektiven Scan Matchings für eine angemessene Ko-Registrierung von 3D- und 2D-Daten, welche mit verschiedenen Sensortypen erfasst wurden, geeignet sind und (v) dass die vorgestellte Strategie zur Erhöhung der Einzigartigkeit von einfachen geome-trischen Merkmalen über die Betrachtung einer optimalen Nachbarschaft für jeden individuellen 3D-Punkt sowie die vorgestellte Strategie zur Selektion einer kom-pakten und robusten Untermenge von relevanten und informativen Merkmalen einen signifikanten, positiven Einfluss auf die Ergebnisse einer 3D-Szenenanalyse haben. Auf diese Weise ermöglichen die entwickelten Verfahren eine effiziente Rekonstruktion und Analyse von großen Bereichen bis auf Stadtgröße und bieten damit großes Potential für zukünftige Forschungsarbeiten.

Note

This book has been developed from a Ph.D. thesis written at the Karlsruhe Institute of Technology (KIT) from 2011 to 2015, further details of which are outlined in the German-language title page, and in the English and German abstracts in the front matter.

Karlsruhe   Martin Weinmann
October 2015

Acknowledgments

First and foremost, I would like to thank Stefan Hinz who accepted to supervise this work from the beginning and provided guidance throughout my time as Ph.D. student. He gave me the great opportunity to carry out investigations in different research directions, to become part of the scientific community and to address my research interests. Under his supervision, I felt all the freedom to develop the core of this work.

Furthermore, I would like to thank Boris Jutzi for his constant advice, the numerous valuable discussions, and his availability throughout my time as Ph.D. student. I greatly appreciate his knowledge and I feel that our cooperation has been inspiring and fruitful.

I would also like to express my sincere gratitude to Uwe Stilla for his interest in my work and for acting as referee of this work despite his heavy workload. Furthermore, I am immensely grateful to Boris Jutzi, Clément Mallet, and Franz Rottensteiner for their helpful comments and relevant remarks on different parts of my work. I also deeply appreciate the valuable discussions with my twin brother Michael Weinmann and his constructive feedback throughout the last years.

I am also very grateful to Nicolas Paparoditis for letting me to be part of his research group at IGN during my stays abroad in France. Furthermore, I would like to thank Clément Mallet who supervised my work at IGN. I greatly appreciate his deep knowledge in numerous research domains, his valuable remarks on my work and the many discussions we had over the last years. I feel that I did learn a lot and hope that our cooperation will continue for a long time. I also want to thank the members of IGN-MATIS for their warm welcome, the interesting discussions, and the many events we have been together.

I would also like to thank my co-authors who contributed to the publications representing fundamental parts of this work. These co-authors are Stefan Hinz, Boris Jutzi, Sven Wursthorn, Jens Leitloff, André Dittrich, Steffen Urban, Uwe Stilla, Ludwig Hoegner, Clément Mallet, Franz Rottensteiner, and Alena Schmidt. Furthermore, I thank the anonymous reviewers of these publications for their effort and their constructive feedback with many relevant remarks.

Moreover, I would like to thank KIT-GRACE, the Graduate School for Climate and Environment at the Karlsruhe Institute of Technology (KIT) for funding my work, my travels to conferences, my participation in summer schools and my first two stays abroad in France in order to support the collaboration between KIT-IPF and IGN-MATIS in 2013. Furthermore, I deeply thank the Karlsruhe House of Young Scientists (KHYS) at KIT for funding my third stay abroad in France in order to support the collaboration between KIT-IPF and IGN-MATIS in 2014.

In particular, I also want to thank Sven Wursthorn for the technical support and Ilse Engelmann for the administrative support throughout the last years.

Of course, I am also very grateful to all my colleagues, ex-colleagues, and friends at KIT-IPF for the comradeship throughout the last years, the lively discussions, and the many events we have been together. Particular thanks in this regard go to Boris Jutzi, Sven Wursthorn, André Dittrich, Steffen Urban, Clémence Dubois, Jens Leitloff, Rosmarie Blomley, Simon Schuffert, Ana Djuricic, Ilse Engelmann, Christian Lucas, Werner Weisbrich, Thomas Vögtle, and Uwe Weidner.

Finally, I am gratefully indebted to my parents Ursula and Robert Weinmann, my brother Frank Weinmann and his wife Tatsiana, and my twin brother Michael Weinmann for their unlimited support during all this time.

Karlsruhe   Martin Weinmann
October 2015

Contents

Chapter 1
Introduction

For us humans, the rich understanding of our world is mainly guided by visual perception which offers a touchless close-range and far-range gathering of information about objects in a scene, their color, their texture, their shape, or their spatial arrangement. This information, in turn, allows to conclude about (i) the number of objects in a scene, (ii) the properties of objects, e.g., with respect to shape, materials, fragility or value, (iii) the distance of objects and their size, (iv) how to interact with the scene, e.g., by grasping an object where thin surface elements might easily break if the object is grasped in a non-adequate way, or (v) how to move through the local environment. Consequently, the imitation of these human capabilities in theory, design, and implementation allows machines to "see" and it is therefore of utmost importance in the fields of photogrammetry, remote sensing, computer vision, and robotics. However, in order to transfer the capabilities of human visual perception to fully automated systems, we may involve several and different sensor types such as digital cameras, range cameras, or laser scanners, and we may even go beyond the human capabilities by for instance involving thermal cameras or multispectral cameras which perceive information in the non-visible spectrum.

Instead of imitating human visual perception, we focus on transferring specific capabilities of human visual perception to fully automated systems and thereby involving different data sources in order to capture different and complementary aspects of real-world objects or scenes. In the scope of this work, we consider those sensor types which provide 3D point cloud data representing the densely sampled and accurately described counterpart of physical object surfaces as most important data source. The nature of such 3D point cloud data requires data structures and processing methods which differ from those used for 2D imagery, yet a representation of a scene in the form of a 3D point cloud provides a sampling of physical object surfaces which merits attraction for a variety of applications. Regarding for instance the acquisition of indoor and outdoor scenes, 3D point cloud data acquired with terrestrial laser scanning systems (e.g., Leica HDS6000 or Riegl LMS-Z360i), mobile laser scanning systems (e.g., L3D2 [1] or Stereopolis II [3]) or range cameras (e.g., PMD[vision] CamCube 2.0, MESA Imaging SR4000 or Microsoft Kinect) are increasingly used for applications reaching from 3D scene reconstruction via geometry processing,

© Springer International Publishing Switzerland 2016
M. Weinmann, *Reconstruction and Analysis of 3D Scenes*,
DOI 10.1007/978-3-319-29246-5_1

shape and appearance modeling, object detection, object recognition, and 3D scene interpretation to large-scale issues. While a framework containing numerous state-of-the art algorithms including filtering, feature estimation, surface reconstruction, registration, model fitting, and segmentation has been released with the Point Cloud Library (PCL) [4] which represents a standalone, large-scale, open project for 2D/3D image and 3D point cloud processing, further improvements toward an advanced 3D point cloud processing are still desirable.

A respective improvement may generally address a variety of different issues amongst which three crucial tasks may be identified. First, it has to be taken into account that, depending on the involved acquisition system, occlusions resulting from objects in the scene may be expected as well as variations in point density which becomes visible in Fig. 1.1 for the visualization of a 3D point cloud acquired with a terrestrial laser scanner. Thus, an adequate digitization and 3D reconstruction of a scene often relies on a combination of several acquired 3D point clouds describing different parts of the considered scene, and appropriate approaches for such a combination of 3D point clouds are required. Second, it might be desirable to map complementary types of data acquired with different sensor types (which typically provide data in the form of images) onto already available 3D point clouds, e.g., in order to obtain a photo-realistic depiction of the considered scene or in order to visualize the heat distribution in the scene. In this regard, for the 3D point cloud depicted in Fig. 1.1, radiometric data could for instance be useful to differentiate between different construction types for buildings as shown in Fig. 1.2. Third, the 3D point cloud depicted in Fig. 1.1 is already sufficient for us humans in order to conclude about specific objects in the scene such as buildings, ground, road inventory, or cars and, hence, it seems desirable to realize such a semantic interpretation of the

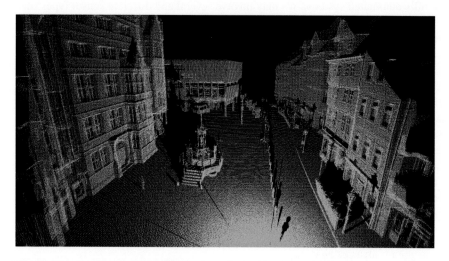

Fig. 1.1 Visualization of a 3D point cloud acquired with a terrestrial laser scanner: occlusions resulting from objects in the scene as well as a varying point density become clearly visible

Fig. 1.2 Visualization of a 3D point cloud acquired with a terrestrial laser scanner and colored with respect to the respective radiometric data

considered scene in a fully automated way. While these three tasks represent fundamental problems in photogrammetry, remote sensing, computer vision, and robotics—and have therefore been addressed in a large number of investigations— respective investigations have mainly focused on a selection of specific subtasks, whereas potential sources for further improvements have not adequately been addressed yet.

1.1 Goals

Generally, in order to understand a scene, we need to reason about the distance and spatial arrangement of objects in the world, gather various properties of these objects and capture the semantic meaning of the world. Accordingly, the understanding of our world via visual perception involves a mapping which, in turn, is composed of three different, but possibly interrelated mappings:

- a *geometric mapping* which maps the real world onto a geometric representation (e.g., a 3D point cloud or a 3D model),
- a *radiometric mapping* which maps the real world onto a radiometric representation (e.g., a color image or a gray-valued image), and
- a *semantic mapping* which maps the real world onto a semantic representation (e.g., a labeling with respect to different objects, different materials, accessible locations, abstract elements, etc.).

Since we focus on transferring capabilities of human visual perception to fully automated systems instead of imitating human visual perception, we may involve several and very different sensors as well as a variety of algorithms in order to obtain an appropriate mapping between the real world and abstract data. Thereby, we have to take into account both the limitations and the interrelations between respective mappings:

- There are sensor types which allow a direct geometric mapping, e.g., laser scanners or range cameras. Furthermore, a geometric mapping may also be derived indirectly by for instance reconstructing a scene from 2D imagery acquired with digital cameras.
- Many sensor types allow a radiometric mapping, e.g., digital cameras, thermal cameras, multispectral cameras, or even modern scanning devices such as laser scanners or range cameras.
- None of the available sensor types directly allows a semantic mapping.
- If we cannot directly obtain a semantic mapping, then we need to exploit the data obtained via a geometric mapping and/or a radiometric mapping in order to allow a semantic reasoning.
- If we cannot directly obtain a geometric mapping, then we need to exploit the data obtained via a radiometric mapping in order to allow a geometric reasoning.
- If we use different sensors for a geometric mapping and/or a radiometric mapping, then we need to correctly align the captured data which is commonly referred to as co-registration of data or data fusion.

The goal of our work consists of the development of novel concepts and methods for the digitization, reconstruction, interpretation, and understanding of static and dynamic indoor and outdoor scenes. Accordingly, we will first focus on the geometric mapping, where we may involve a variety of sensors in order to either directly or indirectly acquire 3D point cloud data representing the measured or derived counterpart of the real world. In this regard, we focus on a complete, dense, and accurate digitization/reconstruction of indoor and outdoor environments and, consequently, it might be advisable to exploit scanning devices in order to directly acquire 3D point cloud data from different viewpoints. As, for each viewpoint, the acquired 3D point cloud data is only measured with respect to the local coordinate frame of the sensor, the data acquired from different viewpoints has to be aligned in a common coordinate frame which is commonly referred to as point cloud registration and addressed in Chap. 4. In this regard, we take into account that most of the currently available scanning devices already provide radiometric information which may significantly facilitate an appropriate alignment. For scene interpretation, we follow the idea of involving further data sources in order to capture different and complementary aspects of real-world objects or scenes, where the additional information is typically represented in the form of 2D imagery. As a consequence, we will have to focus on an adequate co-registration of the acquired data in terms of a mapping of complementary information onto the already available 3D point cloud data which is addressed in Chap. 5. Based on all available data, a subsequent scene interpretation may be carried out in order to derive a semantic mapping and we will address such a semantic reasoning in Chap. 6. Thus, in a summary, the main goals of our work consist of

- a faithful reconstruction of 3D scenes (in both indoor and outdoor environments),
- data fusion in terms of an enrichment of existing 3D point cloud data by adding external data from multiple, but complementary data sources, and
- a semantic reasoning based on all available information.

As input data, we consider (partial) representations of a 3D scene in the form of raw 3D point cloud data consisting of a large number of 3D points as measured counterpart of physical surfaces in the local environment of the sensor, and raw data acquired via complementary data sources. All acquired data is provided to a "black box"—which we will explain in detail in this book—and the derived output data consists of a global reconstruction of the considered scene in the form of a 3D point cloud with correctly aligned radiometric information and respectively assigned point-wise semantic labels.

1.2 Challenges and Main Contributions

As outlined in the previous section, we address the tasks of (i) point cloud registration, (ii) co-registration of 3D point cloud data and complementary information represented in the form of 2D imagery and (iii) 3D scene analysis in the scope of our work. The scientific key contributions presented in this book are:

- **A geometric quantification of the quality of range measurements**:
  In order to assess the quality of scanned 3D points, we consider a 3D point cloud and the respectively derived 2D image representations in the form of range and intensity images. Based on the given range image, we propose two novel image-based measures represented by *range reliability* [8] and *local planarity* [11] in order to assess the quality of scanned 3D points. We explain the chances and limitations of both measures in comparison to other alternatives, and we demonstrate the performance of the proposed measures in the context of filtering raw 3D point cloud data. Furthermore, we demonstrate how to exploit these measures in order to define a consistency check which allows us to filter unreliable feature correspondences between the image representations of different scans and thus to facilitate an efficient and robust registration of 3D point clouds by retaining only those feature correspondences with reliable range information.
- **A shift of paradigms in point cloud registration**:
  Based on our investigations on feature extraction from 2D imagery [6] and 3D point clouds [7], we may conclude that local features are well-suited in order to allow an efficient and robust registration of 3D point clouds. The respective category of registration approaches—which we refer to as *keypoint-based point cloud registration approaches*—has indeed been addressed in literature for years. However, respective investigations mainly focused on the use of specific types of local features representing the most prominent approaches for deriving feature correspondences. Instead, our investigations also take into account that other approaches for extracting local features may even be more promising [5]. Furthermore, recent

investigations typically rely on solving the task of keypoint-based point cloud registration in the classic way by estimating the rigid Euclidean transformation between sets of corresponding 3D points. Instead, we take into account that a rigid transformation only addresses 3D cues in terms of how good corresponding 3D points fit together with respect to the Euclidean distance, whereas feature correspondences between 2D image representations in the form of range and intensity images also allow to involve 2D cues in terms of how good the 2D locations in the image representations fit together. As a result of taking into account both 3D and 2D cues, an improvement in accuracy may be expected for the registration results. Accordingly, we propose to transfer the task of point cloud registration to (i) the task of solving the PnP problem, where we present a projective scan matching which relies on the use of 3D/2D correspondences in order to solve the PnP problem [8, 10–13] and (ii) the task of solving the relative orientation problem, where we present an omnidirectional scan matching which relies on the relative pose estimation based on sets of bearing vectors [5]. Our approaches are principally also suited for point cloud registration in case of continuously moving sensor platforms, yet it is inevitable to introduce slight adaptations [2, 9, 15].

- **An enrichment of existing 3D point cloud data by thermal information**:
  When involving modern scanning devices for data acquisition, the acquired data typically encapsulates both geometric and radiometric information which, in turn, allows an (almost) realistic depiction of a 3D scene. However, the complexity of typical real-world scenes still often hinders important tasks such as object detection or scene interpretation. Consequently, the fusion of data acquired with complementary data sources is desirable as complementary information may be expected to facilitate a variety of tasks. Particularly thermal information offers many advantages for scene analysis, since people may easily be detected as heat sources in typical indoor or outdoor environments and, furthermore, a variety of concealed objects such as heating pipes as well as structural properties such as defects in isolation may be observed. For this reason, we focus on the co-registration of 3D point cloud data and thermal information given in the form of thermal infrared images, and we present an approach relying on a homography estimation for almost planar scenes [14] as well as an approach relying on a projective scan matching technique for general scenes [19].

- **A framework for the semantic interpretation of 3D point cloud data**:
  Once a 3D point cloud is available which adequately represents a considered 3D scene and possibly contains specific attributes for each 3D point in addition to the spatial 3D information, we may be interested in conducting a 3D scene analysis in terms of uniquely assigning each 3D point of the considered 3D point cloud a respective (semantic) class label. While many approaches in this regard have been presented over years, the main effort still typically addresses the classification scheme and seems to be stuck in steadily presenting more and more sophisticated classifiers, although innovative solutions addressing other components in a respective processing workflow might be advisable as well. For this reason, we extensively investigate potential sources for improvements with respect to the classification accuracy and our contributions in this regard are manifold. First, we focus

on assigning each individual 3D point a respective local neighborhood of optimal size in order to increase the distinctiveness of respective geometric features [18, 20]. While optimal neighborhood size selection has still been in an infant stage in the last years, our investigations clearly reveal a significant impact of using locally adaptive 3D neighborhoods instead of standard neighborhood definitions, particularly for our novel approach which we refer to as *eigenentropy-based scale selection* [18, 20], and therefore provide a significant achievement upon which future research may be based. Second, we address the fact that, typically, as many features as possible are exploited in order to compensate a lack of knowledge about the scene and/or data, although it has been proven in practice that irrelevant or redundant information influences the performance of an involved classifier by decreasing its predictive power. Thus, many classifiers are indeed not insensitive to the given dimensionality. In order to overcome such effects, we aim at quantifying the relevance of features and only exploiting suitable features in order to solve the task of 3D scene analysis. Accordingly, we focus on a feature relevance assessment which allows a selection of compact and robust subsets of versatile features [16, 20, 21], and we present a novel general classifier-independent relevance metric for ranking features with respect to their suitability by taking into account different criteria. Third, it is desirable to only involve classifiers which provide accurate classification results while being efficient in terms of processing time and memory consumption. Respective conclusions may only be drawn by an extensive evaluation of different classifiers on a standard benchmark dataset for 3D scene analysis [20]. Finally, we present an extension of our framework toward data-intensive processing [17, 23] as well as an extension allowing the use of contextual information in order to obtain a further improvement of the classification results [22].

All these contributions are based on concepts and methods for feature extraction and, consequently, a deeper understanding on feature design and feature extraction is an important prerequisite. This addresses not only the conceptual ideas behind respective approaches, but also the fundamental motivations arising from current needs. For this purpose, more detailed surveys have been carried out on 2D feature extraction [6] and on 3D feature extraction [7].

1.3 Publications

Most of the material presented in this book has successfully passed a double-blind peer-review process and thus been presented in a series of publications, amongst which are papers for both major conferences and top journals in the field. In the following, these publications are listed according to the respective research topics in order to provide a better overview.

- **Feature extraction**:
  - *M. Weinmann* (2013): Visual features—From early concepts to modern computer vision. In: G. M. Farinella, S. Battiato, and R. Cipolla (Eds.), *Advanced Topics in Computer Vision*. Advances in Computer Vision and Pattern Recognition, Springer, London, UK, pp. 1–34.
  - *M. Weinmann* (2016): Feature extraction from images and point clouds: fundamentals, advances and recent trends. Whittles Publishing, Dunbeath, UK. To appear.

- **Point cloud registration (static sensor platforms)**:
  - *M. Weinmann*, Mi. Weinmann, S. Hinz, and B. Jutzi (2011): Fast and automatic image-based registration of TLS data. In: F. Bretar, W. Wagner, and N. Paparoditis (Eds.), Advances in LiDAR Data Processing and Applications. *ISPRS Journal of Photogrammetry and Remote Sensing*, Vol. 66 (6), Supplement, pp. S62–S70.
  - *M. Weinmann* and B. Jutzi (2011): Fully automatic image-based registration of unorganized TLS data. In: D. D. Lichti and A. F. Habib (Eds.), ISPRS Workshop Laser Scanning 2011. *The International Archives of the Photogrammetry, Remote Sensing and Spatial Information Sciences*, Vol. XXXVIII-5/W12, pp. 55–60.
  - *M. Weinmann*, S. Wursthorn, and B. Jutzi (2011): Semi-automatic image-based fusion of range imaging data with different characteristics. In: U. Stilla, F. Rottensteiner, H. Mayer, B. Jutzi, and M. Butenuth (Eds.), PIA11: Photogrammetric Image Analysis 2011. *The International Archives of the Photogrammetry, Remote Sensing and Spatial Information Sciences*, Vol. XXXVIII-3/W22, pp. 119–124.
  - *M. Weinmann* and B. Jutzi (2013): Fast and accurate point cloud registration by exploiting inverse cumulative histograms (ICHs). *Proceedings of the IEEE Joint Urban Remote Sensing Event*, pp. 218–221.
  - *M. Weinmann* and B. Jutzi (2015): Geometric point quality assessment for the automated, markerless and robust registration of unordered TLS point clouds. In: C. Mallet, N. Paparoditis, I. Dowman, S. Oude Elberink, A.-M. Raimond, F. Rottensteiner, M. Yang, S. Christophe, A. Çöltekin, and M. Brédif (Eds.), ISPRS Geospatial Week 2015. *ISPRS Annals of the Photogrammetry, Remote Sensing and Spatial Information Sciences*, Vol. II-3/W5, pp. 89–96.
  - S. Urban and *M. Weinmann* (2015): Finding a good feature detector-descriptor combination for the 2D keypoint-based registration of TLS point clouds. In: C. Mallet, N. Paparoditis, I. Dowman, S. Oude Elberink, A.-M. Raimond, F. Rottensteiner, M. Yang, S. Christophe, A. Çöltekin, and M. Brédif (Eds.), ISPRS Geospatial Week 2015. *ISPRS Annals of the Photogrammetry, Remote Sensing and Spatial Information Sciences*, Vol. II-3/W5, pp. 121–128.

- **Point cloud registration (moving sensor platforms)**:
  - S. Hinz, *M. Weinmann*, P. Runge, and B. Jutzi (2011): Potentials of image based active ranging to capture dynamic scenes. In: C. Heipke, K. Jacobsen, F. Rottensteiner, S. Müller, and U. Sörgel (Eds.), ISPRS Hannover Workshop 2011: High-Resolution Earth Imaging for Geospatial Information. *The International Archives of the Photogrammetry, Remote Sensing and Spatial Information Sciences*, Vol. XXXVIII-4/W19, pp. 143–147.
  - *M. Weinmann* and B. Jutzi (2012): A step towards dynamic scene analysis with active multi-view range imaging systems. In: M. Shortis, N. Paparoditis, and C. Mallet (Eds.), XXII ISPRS Congress, Technical Commission III. *The International Archives of the Photogrammetry, Remote Sensing and Spatial Information Sciences*, Vol. XXXIX-B3, pp. 433–438.
  - *M. Weinmann*, A. Dittrich, S. Hinz, and B. Jutzi (2013): Automatic feature-based point cloud registration for a moving sensor platform. In: C. Heipke, K. Jacobsen, F. Rottensteiner, and U. Sörgel (Eds.), ISPRS Hannover Workshop 2013: High-Resolution Earth Imaging for Geospatial Information. *The International Archives of the Photogrammetry, Remote Sensing and Spatial Information Sciences*, Vol. XL-1/W1, pp. 373–378.

- **Co-registration of 3D point cloud data and thermal information**:
  - *M. Weinmann*, L. Hoegner, J. Leitloff, U. Stilla, S. Hinz, and B. Jutzi (2012): Fusing passive and active sensed images to gain infrared-textured 3D models. In: M. Shortis and N. El-Sheimy (Eds.), XXII ISPRS Congress, Technical Commission I. *The International Archives of the Photogrammetry, Remote Sensing and Spatial Information Sciences*, Vol. XXXIX-B1, pp. 71–76.
  - *M. Weinmann*, J. Leitloff, L. Hoegner, B. Jutzi, U. Stilla, and S. Hinz (2014): Thermal 3D mapping for object detection in dynamic scenes. In: C. Toth and B. Jutzi (Eds.), ISPRS Technical Commission I Symposium. *ISPRS Annals of the Photogrammetry, Remote Sensing and Spatial Information Sciences*, Vol. II-1, pp. 53–60.

- **3D scene analysis**:
  - *M. Weinmann*, B. Jutzi, and C. Mallet (2013): Feature relevance assessment for the semantic interpretation of 3D point cloud data. In: M. Scaioni, R. C. Lindenbergh, S. Oude Elberink, D. Schneider, and F. Pirotti (Eds.), ISPRS Workshop Laser Scanning 2013. *ISPRS Annals of the Photogrammetry, Remote Sensing and Spatial Information Sciences*, Vol. II-5/W2, pp. 313–318.
  - *M. Weinmann*, B. Jutzi, and C. Mallet (2014): Describing Paris: automated 3D scene analysis via distinctive low-level geometric features. *Proceedings of the IQmulus Workshop on Processing Large Geospatial Data*, pp. 1–8.
  - *M. Weinmann*, B. Jutzi, and C. Mallet (2014): Semantic 3D scene interpretation: a framework combining optimal neighborhood size selection with relevant features. In: K. Schindler and N. Paparoditis (Eds.), ISPRS Technical Commission

III Symposium. *ISPRS Annals of the Photogrammetry, Remote Sensing and Spatial Information Sciences*, Vol. II-3, pp. 181–188.

- *M. Weinmann*, S. Urban, S. Hinz, B. Jutzi, and C. Mallet (2015): Distinctive 2D and 3D features for automated large-scale scene analysis in urban areas. *Computers & Graphics*, Vol. 49, pp. 47–57.
- *M. Weinmann*, B. Jutzi, S. Hinz, and C. Mallet (2015): Semantic point cloud interpretation based on optimal neighborhoods, relevant features and efficient classifiers. In: K. Schindler (Ed.), Photogrammetric Computer Vision 2014—Best Papers of the ISPRS Technical Commission III Symposium. *ISPRS Journal of Photogrammetry and Remote Sensing*, Vol. 105, pp. 286–304.
- *M. Weinmann*, A. Schmidt, C. Mallet, S. Hinz, F. Rottensteiner, and B. Jutzi (2015): Contextual classification of point cloud data by exploiting individual 3D neighborhoods. In: U. Stilla and C. Heipke (Eds.), PIA15+HRIGI15—Joint ISPRS Conference. *ISPRS Annals of the Photogrammetry, Remote Sensing and Spatial Information Sciences*, Vol. II-3/W4, pp. 271–278.
- *M. Weinmann*, C. Mallet, S. Hinz, and B. Jutzi (2015): Efficient interpretation of 3D point clouds by assessing feature relevance. *AVN–Allgemeine Vermessungs-Nachrichten*, Vol. 10/2015, pp. 308–315.

The close connection of the material presented in this book to other research topics also allowed a participation in different projects, and respective results have been presented in further publications which are not part of this book:

- L. Hoegner, *M. Weinmann*, B. Jutzi, S. Hinz, and U. Stilla (2013): Co-registration of time-of-flight (TOF) camera generated 3D point clouds and thermal infrared images (IR). In: E. Seyfert (Ed.), *Tagungsband der 33. Wissenschaftlich-Technischen Jahrestagung der DGPF*, Vol. 22, pp. 481–488.
- L. Hoegner, *M. Weinmann*, B. Jutzi, S. Hinz, and U. Stilla (2013): Synchrone Koregistrierung von 3D Punktwolken und thermischen Infrarotbildern. In: T. Luhmann and C. Müller (Eds.), *Photogrammetrie – Laserscanning – Optische 3D-Messtechnik: Beiträge der 12. Oldenburger 3D-Tage 2013*. Wichmann Verlag, Heidelberg, Germany.
- L. Hoegner, L. Roth, *M. Weinmann*, B. Jutzi, S. Hinz, and U. Stilla (2014): Fusion von Time-of-Flight Entfernungsdaten und thermalen IR-Bildern. *AVN–Allgemeine Vermessungs-Nachrichten*, Vol. 5/2014, pp. 192–197.
- L. Hoegner, A. Hanel, *M. Weinmann*, B. Jutzi, S. Hinz, and U. Stilla (2014): Towards people detection from fused time-of-flight and thermal infrared images. In: K. Schindler and N. Paparoditis (Eds.), ISPRS Technical Commission III Symposium. *The International Archives of the Photogrammetry, Remote Sensing and Spatial Information Sciences*, Vol. XL-3, pp. 121–126.
- B. Jutzi, *M. Weinmann*, and J. Meidow (2013): Improved UAV-borne 3D mapping by fusing optical and laserscanner data. In: G. Grenzdörffer and R. Bill (Eds.), UAV-g2013. *The International Archives of the Photogrammetry, Remote Sensing and Spatial Information Sciences*, Vol. XL-1/W2, pp. 223–228.

- B. Jutzi, *M. Weinmann*, and J. Meidow (2014): Weighted data fusion for UAV-borne 3D mapping with camera and line laserscanner. *International Journal of Image and Data Fusion*, Vol. 5 (3), pp. 226–243.
- A. Djuricic, *M. Weinmann*, and B. Jutzi (2014): Potentials of small, lightweight and low cost multi-echo laser scanners for detecting grape berries. In: F. Remondino and F. Menna (Eds.), ISPRS Technical Commission V Symposium. *The International Archives of the Photogrammetry, Remote Sensing and Spatial Information Sciences*, Vol. XL-5, pp. 211–216.
- R. Blomley, *M. Weinmann*, J. Leitloff, and B. Jutzi (2014): Shape distribution features for point cloud analysis—A geometric histogram approach on multiple scales. In: K. Schindler and N. Paparoditis (Eds.), ISPRS Technical Commission III Symposium. *ISPRS Annals of the Photogrammetry, Remote Sensing and Spatial Information Sciences*, Vol. II-3, pp. 9–16.
- A. C. Braun, *M. Weinmann*, S. Keller, R. Müller, P. Reinartz and S. Hinz (2015): The EnMAP contest: developing and comparing classification approaches for the Environmental Mapping and Analysis Programme—Dataset and first results. In: C. Mallet, N. Paparoditis, I. Dowman, S. Oude Elberink, A.-M. Raimond, G. Sithole, G. Rabatel, F. Rottensteiner, X. Briottet, S. Christophe, A. Çöltekin, and G. Patanè (Eds.), ISPRS Geospatial Week 2015. *The International Archives of the Photogrammetry, Remote Sensing and Spatial Information Sciences*, Vol. XL-3/W3, pp. 169–175.

1.4 The Proposed Framework

In the scope of this work, we present a novel framework for advanced 3D point cloud processing from raw data to semantic objects. This framework consists of several components as shown in Fig. 1.3 and addresses

- a preprocessing of acquired data, e.g., in terms of distinguishing between reliable and unreliable measurements and a respective filtering of raw 3D point cloud data,
- feature extraction delivering the fundamental requirements for all subsequent components,
- a geometric enrichment via point cloud registration, i.e., the alignment of several 3D point clouds in a common coordinate frame,
- a radiometric enrichment via co-registration of 3D and 2D data, e.g., in terms of thermal 3D mapping which aims at mapping thermal information represented in the form of thermal infrared images onto acquired 3D point cloud data, and
- a semantic enrichment via 3D scene analysis in terms of assigning a semantic class label to each 3D point of given 3D point cloud data.

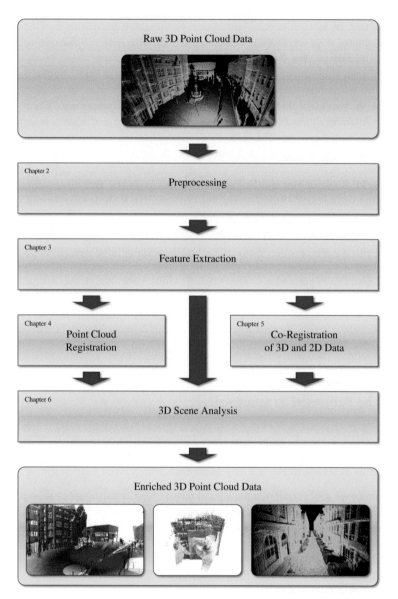

Fig. 1.3 The proposed framework for advanced 3D point cloud processing from raw data to semantic objects. As indicated, the single components are addressed in detail in the following chapters. In the scope of our work, we focus on a semantic enrichment of 3D point cloud data via 3D scene analysis, but this may also rely on a geometric enrichment via point cloud registration and a radiometric enrichment via co-registration of 3D and 2D data

1.5 Book Outline

This book is structured in seven chapters and addresses the reconstruction and analysis of 3D scenes, whereby the focus is put on the presentation of an end-to-end processing workflow from raw 3D point cloud data to semantic objects in the scene.

In this chapter, we have provided an introduction which focuses on explaining the goals of our work, the main challenges when addressing these goals and the scientific contributions developed in this context. We have additionally provided a list of our publications, taken a glance view on our novel framework for advanced 3D point cloud processing from raw data to semantic objects and finally given an overview on the topics presented in the whole book.

In Chap. 2, we consider fundamentals in the form of general definitions that will be important for a consideration of our work in the global context. Since, in this book, we focus on the use of 3D point cloud data as input for our methodology, we briefly explain how such 3D point cloud data may be acquired indirectly from 2D imagery via passive techniques and how it may be acquired directly by involving active techniques. Particularly the active techniques offer many advantages and we therefore consider respective systems for data acquisition in the scope of this book. An analysis of such acquisition systems reveals that the acquired 3D point cloud data may easily be represented in the form of range and intensity images which, in turn, might facilitate a variety of applications in terms of simplicity and efficiency. In this regard, we consider point quality assessment where the aim is to quantify the quality of the single range measurements captured with the involved acquisition system, and we present and discuss two novel approaches and their potential in comparison to related work.

In Chap. 3, we briefly explain fundamental concepts for extracting features from 2D imagery and 3D point cloud data. In this regard, we derive a definition for the general term of a *feature* and consider different types of features that might occur in 2D imagery or 3D point cloud data. These fundamentals have become the basis and core of a plethora of applications in photogrammetry, remote sensing, computer vision, and robotics, and they also provide the basis for the subsequent chapters.

In Chap. 4, we focus on transforming 3D point clouds—where each 3D point cloud has been acquired with respect to the local coordinate frame of the acquisition system—into a common coordinate frame. We consider related work and thereby particularly focus on efficient approaches relying on the use of keypoints. Based on these considerations, we present a novel framework for point cloud registration which exploits the 2D representations of acquired 3D point cloud data in the form of range and intensity images in order to quantify the quality of respective range measurements and subsequently reliably extract corresponding information between different scans. In this regard, we may consider corresponding information in the respective image representations or in 3D space, and we may also introduce a weighting of derived correspondences with respect to different criteria. Thus, the correspondences may serve as input for the registration procedure for which we present two novel approaches. An evaluation on a benchmark dataset clearly reveals the performance of our framework, and we discuss the derived results with respect to different aspects including registration accuracy and computational efficiency.

In Chap. 5, we focus on the enrichment of acquired 3D point clouds by additional information acquired with a thermal camera. We reflect related work and present a novel framework for thermal 3D mapping. Our framework involves a radiometric correction, a geometric calibration, feature extraction and matching, and two approaches for the co-registration of 3D and 2D data. For the example of an indoor scene, we demonstrate the performance of our framework in terms of both accuracy and applicability, and we discuss the derived results with respect to potential use-cases.

In Chap. 6, we focus on 3D scene analysis where the aim is to uniquely assign each 3D point of a given 3D point cloud a respective (semantic) class label. We provide a survey on related work and present a novel framework addressing neighborhood selection, feature extraction, feature selection, and classification. Furthermore, we consider extensions of our framework toward large-scale capability and toward the use of contextual information. In order to demonstrate the performance of our framework, we provide an extensive experimental evaluation on publicly available benchmark datasets and discuss the derived results with respect to classification accuracy, computational efficiency and applicability of involved methods.

In Chap. 7, we summarize the contents presented in this book and provide concluding remarks as well as suggestions for future work.

References

1. Goulette F, Nashashibi F, Abuhadrous I, Ammoun S, Laurgeau C (2006) An integrated on-board laser range sensing system for on-the-way city and road modelling. Int Arch Photogramm Remote Sens Spat Inf Sci XXXVI-1:1–6
2. Hinz S, Weinmann M, Runge P, Jutzi B (2011) Potentials of image based active ranging to capture dynamic scenes. Int Arch Photogramm Remote Sens Spat Inf Sci XXXVIII-4/W19:143–147
3. Paparoditis N, Papelard J-P, Cannelle B, Devaux A, Soheilian B, David N, Houzay E (2012) Stereopolis II: a multi-purpose and multi-sensor 3D mobile mapping system for street visualisation and 3D metrology. Revue Française de Photogrammétrie et de Télédétection 200:69–79
4. Rusu RB, Cousins S (2011) 3D is here: Point Cloud Library (PCL). In: Proceedings of the IEEE international conference on robotics and automation, pp 1–4
5. Urban S, Weinmann M (2015) Finding a good feature detector-descriptor combination for the 2D keypoint-based registration of TLS point clouds. ISPRS Ann Photogramm Remote Sens Spat Inf Sci II-3/W5:121–128
6. Weinmann M (2013) Visual features—From early concepts to modern computer vision. In: Farinella GM, Battiato S, Cipolla R (eds) Advanced topics in computer vision. Advances in computer vision and pattern recognition. Springer, London, pp 1–34
7. Weinmann M (2016) Feature extraction from images and point clouds: fundamentals, advances and recent trends. Whittles Publishing, Dunbeath (to appear)
8. Weinmann M, Jutzi B (2011) Fully automatic image-based registration of unorganized TLS data. Int Arch Photogramm Remote Sens Spat Inf Sci XXXVIII-5/W12:55–60
9. Weinmann M, Jutzi B (2012) A step towards dynamic scene analysis with active multi-view range imaging systems. Int Arch Photogramm Remote Sens Spat Inf Sci XXXIX-B3:433–438

10. Weinmann M, Jutzi B (2013) Fast and accurate point cloud registration by exploiting inverse cumulative histograms (ICHs). In: Proceedings of the joint urban remote sensing event (JURSE), pp 218–221

11. Weinmann M, Jutzi B (2015) Geometric point quality assessment for the automated, markerless and robust registration of unordered TLS point clouds. ISPRS Ann Photogramm Remote Sens Spat Inf Sci II-3/W5:89–96

12. Weinmann M, Weinmann Mi, Hinz S, Jutzi B (2011) Fast and automatic image-based registration of TLS data. ISPRS J Photogramm Remote Sens 66(6):S62–S70

13. Weinmann M, Wursthorn S, Jutzi B (2011) Semi-automatic image-based fusion of range imaging data with different characteristics. Int Arch Photogramm Remote Sens Spat Inf Sci XXXVIII-3/W22:119–124

14. Weinmann M, Hoegner L, Leitloff J, Stilla U, Hinz S, Jutzi B (2012) Fusing passive and active sensed images to gain infrared-textured 3D models. Int Arch Photogramm Remote Sens Spat Inf Sci XXXIX-B1:71–76

15. Weinmann M, Dittrich A, Hinz S, Jutzi B (2013) Automatic feature-based point cloud registration for a moving sensor platform. Int Arch Photogramm Remote Sens Spat Inf Sci XL-1/W1:373–378

16. Weinmann M, Jutzi B, Mallet C (2013) Feature relevance assessment for the semantic interpretation of 3D point cloud data. ISPRS Ann Photogramm Remote Sens Spat Inf Sci II-5/W2:313–318

17. Weinmann M, Jutzi B, Mallet C (2014) Describing Paris: automated 3D scene analysis via distinctive low-level geometric features. In: Proceedings of the IQmulus workshop on processing large geospatial data, pp 1–8

18. Weinmann M, Jutzi B, Mallet C (2014): Semantic 3D scene interpretation: a framework combining optimal neighborhood size selection with relevant features. ISPRS Ann Photogramm Remote Sens Spat Inf Sci II-3:181–188

19. Weinmann M, Leitloff J, Hoegner L, Jutzi B, Stilla U, Hinz S (2014) Thermal 3D mapping for object detection in dynamic scenes. ISPRS Ann Photogramm Remote Sens Spat Inf Sci II-1:53–60

20. Weinmann M, Jutzi B, Hinz S, Mallet C (2015) Semantic point cloud interpretation based on optimal neighborhoods, relevant features and efficient classifiers. ISPRS J Photogramm Remote Sens 105:286–304

21. Weinmann M, Mallet C, Hinz S, Jutzi B (2015) Efficient interpretation of 3D point clouds by assessing feature relevance. AVN—Allg Vermess-Nachr 10(2015):308–315

22. Weinmann M, Schmidt A, Mallet C, Hinz S, Rottensteiner F, Jutzi B (2015) Contextual classification of point cloud data by exploiting individual 3D neighborhoods. ISPRS Ann Photogramm Remote Sens Spat Inf Sci II-3/W4:271–278

23. Weinmann M, Urban S, Hinz S, Jutzi B, Mallet C (2015) Distinctive 2D and 3D features for automated large-scale scene analysis in urban areas. Comput Graph 49:47–57

Chapter 2
Preliminaries of 3D Point Cloud Processing

In this chapter, we provide preliminaries of 3D point cloud processing. For this purpose, we first describe the general concept of scene acquisition (Sect. 2.1) and subsequently provide basic definitions for the most important terms which are important to consider the contents presented in this book in the global context (Sect. 2.2). Based on these explanations, we focus on the use of 3D point cloud data and explain how 3D point cloud data may be acquired (Sect. 2.3) and how 3D point cloud data may be transformed to respective 2D image representations in the form of range images and intensity images (Sect. 2.4). Since acquired 3D point cloud data is typically corrupted with a certain amount of noise, we enlighten the issue of filtering raw 3D point cloud data in terms of removing unreliable 3D points (Sect. 2.5). In this regard, we present two novel measures for point quality assessment and discuss their chances and limitations in detail. Finally, we provide concluding remarks (Sect. 2.6).

2.1 From the Real World to a Scene and Its Representation

In order to transfer specific capabilities of human visual perception to fully automated systems, the main goal consists of gathering information about the *real world* for which we may generally involve different sensor types. However, similar to human vision, the sensor types typically reveal a limited field-of-view and partially also a limited perception in case of distant objects. Thus, it is impossible to describe the real world as a whole but only a small part, and we will therefore define a *scene* as the considered part of the real world, i.e., a collection of shapes and reflectance [39].

Considering a scene with a specific sensor type (Fig. 2.1), the sensor type establishes a function for mapping the considered scene onto *data*. In this regard, the use of different sensor types (e.g., digital cameras, thermal cameras, multispectral cameras, range cameras, laser scanners, etc.) typically results in data in the form of 2D imagery or 3D point cloud data. In the context of decision and control tasks, however, the measured data generally has no information since it is only measured data, whereas *information* reduces the uncertainty with respect to a specific *task* or

Fig. 2.1 Generalized relation between scene, data, data representations, and task: the scene (e.g., a tree) is mapped onto data (e.g., 3D point cloud data) which, in turn, is transformed into a different data representation (e.g., specific features) suited to solve a specific task (e.g., 3D scene analysis)

an *application*. In the following chapters, we will assume that the measured data contains information which might help us to reduce the uncertainty about a specific task, e.g., to estimate the viewpoint or to recognize objects.

The acquired data in the form of 2D imagery or 3D point cloud data is however typically not suited for solving specific tasks directly and, consequently, the data has to be transferred to a different *data representation* being a specific description and thus a function of the data. This means that a transformation of data from one space to another encoding is involved, whereby the transformation is described via a *function* which, in turn, may be based on a model, some lines of code or an algorithm, and typically depends on the respective task. For instance, it may be desirable to try to take away as much as possible from the data and retain as much as matters for the task, i.e., to preserve only the information which is relevant for a specific task. Note that this is in accordance with the idea to keep all the data if we do not know the task. Furthermore, we may extend these considerations by introducing a *memory* as a mechanism to store data or data representations and *knowledge* as something useful one may draw conclusions from.

Note that the applied definition of *information* follows seminal work [39, 40] which is in contrast to traditional information theory based on fundamental ideas of Claude Elwood Shannon and Norbert Wiener. Traditional information theory considers information as "complexity of the data, regardless of its use and regardless of the nuisance factors affecting it," whereby the complexity of the data is often described with the entropy of a respective distribution and nuisance factors may for instance be represented by contrast, viewpoint, illumination, occlusions, quantization, and noise [40]. With increasing complexity of the data, it becomes more and more costly to store and transmit the data. Thus, the traditional definition of information is tailored to data storage/compression and transmission tasks, where the semantic aspect of the data is irrelevant[1] [40]. However, if we intend to use the data for a different task, a different definition may be desirable which—as for instance stated in [39]—also accounts for questions regarding the information an image contains about the scene it portrays or the value of an image for recognizing objects in the scene or navigating through the scene. Consequently, the definition of information should rather

---

[1]Note that, in the context of data storage/compression and transmission tasks, the most costly signal is represented by white noise [40].

address the part of the data that matters for the task and thus be defined relative to the scene. This, in turn, is in accordance with biological systems that perceive stimuli and perform actions or make decisions in a way that maximizes some decision or control objective [40], e.g., in order to intelligently interact with the surrounding. For this reason, we follow the ideas presented in [39, 40] and consider the notion of information in the context of performing a decision or action based on sensory data, whereby a decision may represent any task of detection, localization, categorization, and recognition of objects [40]. By focusing on the definition of information as the part of the data that matters for the task, the complexity of the data is not relevant to decision and control tasks[2] [40]. For more detailed and further considerations, we refer to [40].

2.2 On Points and Clouds, and Point Clouds

In this section, we intend to provide basic definitions for the most important terms used across the whole book:

- The term *point* is used as in geometry, where it specifies a unique location in a specific space. In the scope of our work, we consider the Euclidean space $\mathbb{R}^D$ with $D = 2$ for points in 2D imagery such as image locations or 2D keypoints and $D = 3$ for points in 3D space such as scanned 3D points, reconstructed 3D points, or 3D keypoints. Hence, a point has no dimensions such as length, area, or volume.
- While there is certainly a meteorological association, we refer the term *cloud* to cloud computing and thus a complex interplay of application systems which are executed externally (within the cloud) and operated via specific interfaces (mostly via internet platforms). In this regard, different applications focus on (i) the external processing of computationally intensive methods in terms of runtime and memory and (ii) the transfer of results to smartphones, tablets, notebooks, desktop computers, or servers, e.g., in the form of web services.
- The term *point cloud* is used to describe a set of data points in a given space. Following [32], we may consider a point cloud as a data structure used to represent a collection of multidimensional points. As our focus is clearly set on 3D point clouds representing the measured or generated counterpart of physical surfaces in a scene, we will concentrate on 3D point clouds as a collection of 3D points which, in turn, are characterized by spatial *XYZ*-coordinates and may optionally be assigned additional attributes such as intensity information, thermal information, specific properties (e.g., in terms of orientation and scale), or any abstract information.

Even though it is intended that some of the developed concepts and methods are integrated in a web service, we only focus on data processing on notebooks, desktop computers, or servers without cloud computing. This is in analogy to a variety of

---

[2]Note that extremely complex data may be useless for a decision, while extremely simple data could directly be relevant for the same decision [40].

research domains where the development of efficient methods with respect to process-ing time, memory consumption, and financial costs is in the focus. For instance, the 3D reconstruction of large parts of the city of Rome, Italy, has been conducted based on 150 k images in less than a day on a huge cluster with almost 500 cores [1]—which we may refer to as a data processing on a *cloudy day* due to the involved cloud computing—whereas the processing of large photo collections within the span of a day on a single PC may be referred to as a processing on a *cloudless day* [13] which is certainly favorable considering the related financial costs.

2.3 Point Cloud Acquisition

As all the concepts and methods developed in the scope of our work are based on the use of 3D point cloud data as input, we will briefly discuss the acquisition of such 3D point cloud data. Generally, the acquisition of 3D point clouds representing the counterpart of physical object surfaces in a scene has been widely investigated over the last decades. While a variety of acquisition systems based on different principles may be involved, particularly the optical methods have proven to be favorable as they offer an efficient, touchless, and even far-range acquisition of 3D structures in both indoor and outdoor environments—from small scales such as single rooms up to large scales such as city scale or even beyond—and, thus, they have gained much importance in photogrammetry, remote sensing, computer vision, and robotics.

Generally, we may categorize the different techniques proposed for optical 3D shape acquisition with respect to the measurement principle they are based on, and numerous valuable surveys following this categorization may be found in literature [10, 20, 30, 34, 57]. Basically, this categorization first focuses on the distinction between passive and active techniques. While *passive techniques* are used for scenes with reasonable ambient lighting conditions and only collect information, *active techniques* focus on actively manipulating the observed scene, e.g., by projecting structured light patterns in the form of stripe patterns or point patterns via emitting electromagnetic radiation in either the visible spectrum or in the form of laser light (typically in the infrared (IR) spectrum). Based on this more general distinction, a further distinction may directly refer to the spatial arrangement of viewpoints and the positions of involved illumination sources. In this regard, passive techniques may be based on a single view or on multiple views of a scene. For active techniques, one may furthermore distinguish between a monostatic sensor configuration, where the emitter component and the receiver component are at the same location, or a multistatic configuration, where the emitter component and the receiver component are spatially separated and even several of such components may be involved. From these distinctions, a taxonomy of optical 3D shape acquisition techniques as illus-trated in Fig. 2.2 may be derived. In order to understand the main principles and the basis for the subsequent chapters, we provide a short description of the most important ideas in the following subsections. For an exhaustive survey, we refer to the aforementioned literature. However, we want to point out that, depending on the

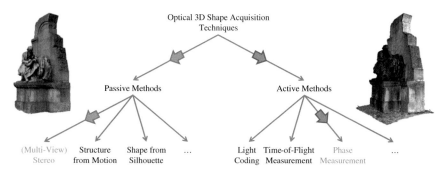

Fig. 2.2 Taxonomy of optical 3D shape acquisition techniques, an exemplary point cloud derived via Multi-View Stereo (MVS) techniques (*left*) and an exemplary point cloud derived via phase-based measurements (*right*)

involved acquisition technique and the used device, the acquired 3D point cloud data may be corrupted with more or less noise and, in addition to spatial 3D information in the form of XYZ-coordinates, respective point attributes such as color or intensity information may be acquired as well. Furthermore, it should be taken into account that—despite a variety of optical 3D shape acquisition techniques—shape reconstruction still remains challenging for surfaces which exhibit a complex surface reflectance behavior such as highly specular object surfaces as given for mirroring objects where an adequate 3D reconstruction may, for instance, be achieved via more sophisticated techniques [56].

2.3.1 Passive Techniques

Passive techniques for point cloud generation only rely on radiometric information represented in the form of 2D imagery and thus simple intensity measurements per pixel. In order to reconstruct 3D structures from such intensity images, different approaches have been presented and we will focus on the two most commonly applied strategies.

The first strategy is followed by *stereo matching techniques* which exploit two or more images of a scene, acquired with either multiple cameras or one moving camera, and estimate the respective 3D structure of the scene by finding corresponding points in the intensity images and converting their 2D locations into 3D depth values based on triangulation [36, 46]. Classical stereo matching techniques are based on the use of two images due to early findings in human visual perception according to which the depth of a scene is perceived based on the differences in appearance between what is seen by the left eye and the right eye. In contrast, Multi-View Stereo (MVS) techniques focus on reconstructing a complete 3D object model from a collection of images taken from known viewpoints [37].

The second strategy is followed by *Structure-from-Motion (SfM) techniques* which focus on the simultaneous recovery of both the 3D structure of a scene and the camera pose from a large number of point tracks, i.e., corresponding image locations [46]. One of the most general approaches to SfM has been presented with *Bundle Adjustment* [47] which is able to simultaneously recover 3D structures, camera pose, and even the intrinsic camera parameters via the consideration of bundles of rays connecting the camera centers to 3D scene points and an adjustment in terms of iteratively minimizing the reprojection error.

2.3.2   Active Techniques

Whereas passive techniques only rely on information arising from simple intensity measurements per pixel, active techniques are based on actively manipulating the observed scene by involving a scanning device which emits a signal and records information arising from respective observations in the scene. In this context, the emitted signal may either be a coded structured light pattern in the visible or infrared spectrum, or electromagnetic radiation in the form of modulated laser light. For both cases, we briefly reflect the basic ideas in order to understand the measurement principles of the scanning devices used in the following chapters.

Scanning devices exploiting *structured light projection* actively manipulate the observed scene by projecting a coded structured light pattern, and thus manipulating the illumination of the scene. Via the projected pattern, particular labels are assigned to 3D scene points which, in turn, may easily be decoded in images when imaging the scene and the projected pattern with a camera. Accordingly, such synthetically generated features allow to robustly establish feature correspondences, and the respective 3D coordinates may easily and reliably be recovered by triangulation. Generally, techniques based on the use of structured light patterns may be classified depending on the pattern codification strategy [10, 33]:

- Direct codification strategies exploit a pattern which is typically based on a large range of either gray or color values. Thus, each pixel may be labeled with the respective gray or color value of the pattern at this pixel.
- Time-multiplexing strategies exploit a temporal coding, where a sequence of structured light patterns is projected onto the scene. Thus, each pixel may be assigned a codeword consisting of its illumination value across the projected patterns. The respective patterns may, for instance, be based on binary codes or Gray codes and phase shifting.
- Spatial neighborhood codification strategies exploit a unique pattern. The label associated to a pixel is derived from the spatial pattern distribution within its local neighborhood. Thus, labels of neighboring pixels share information and provide an interdependent coding.

Representing one of the most popular devices based on structured light projection, the Microsoft Kinect exploits an RGB camera, an IR camera, and an IR projector.

The IR projector projects a known structured light pattern in the form of a random dot pattern onto the scene. As IR camera and IR projector form a stereo pair, the pattern matching in the IR image results in a raw disparity image which, in turn, is read out in the form of an 11-bit depth image. For respective investigations on the accuracy and capabilities of Microsoft Kinect, we refer to [25, 38, 53, 59].

A different strategy for measuring the distance between a scanning device and object surfaces in the local environment is followed by exploiting *Time-of-Flight (ToF) or phase measurements* which are typically exploited by terrestrial or mobile laser scanners and by Time-of-Flight (ToF) cameras. According to seminal work [21, 24, 44], respective scanning devices may be categorized with respect to laser type, modulation technique, measurement principle, detection technique, or configuration between emitting and receiving component of the device. Generally, such scanning devices illuminate a scene with modulated laser light and analyze the backscattered signal. More specifically, laser light is emitted by the scanning device and transmitted to an object. At the object surface, the laser light is partially reflected and, finally, a certain amount of the laser light reaches the receiver unit of the scanning device. The measurement principle is therefore of great importance as it may be based on different signal properties such as amplitude, frequency, polarization, time, or phase. Many scanning devices are based on measuring the time t between emitting and receiving a laser pulse, i.e., the respective *time-of-flight*, and exploiting the measured time t in order to derive the distance r between the scanning device and the respective 3D scene point. Alternatively, a range measurement r may be derived from phase information by exploiting the phase difference $\Delta\phi$ between emitted and received signal. However, the phase difference $\Delta\phi$ only is a wrapped phase and thus the corresponding range is ambiguous so that multiples of 2π have to be added in order to recover the unwrapped phase ϕ and an appropriate range measurement r. For such a disambiguation, various image-based or hardware-based phase unwrapping procedures have been proposed [12, 22, 23]. In order to get from single 3D scene points to the geometry of object surfaces, respective scanning devices are typically mounted on a rotating platform[3] which, in turn, allows a sequential scanning of the scene by successively measuring distances for discrete points on a given scan raster. Due to the sequential scanning, however, respective scanning devices are not suited for the acquisition of dynamic scenes. In contrast, modern range cameras use a sensor array which allows to simultaneously acquire range information for all points on the considered scan raster and thus estimate the scene geometry in a single shot. Such range cameras meanwhile allow to acquire depth maps at relatively high frame rates of about 25 fps [10, 26, 31, 57] and are thus also applicable in order to capture dynamic scenes.

---

[3]In this context, a rotation around a single rotation axis is for instance applied for line laser scanners which thus provide measurements on a 1D scan grid, whereas standard terrestrial laser scanners involve a rotation in both horizontal and vertical direction in order to provide measurements on a 2D scan grid.

2.4 Generation of 2D Image Representations for 3D Point Clouds

In the scope of our work, we focus on point clouds acquired with (terrestrial or mobile) laser scanners and range cameras. Thereby, laser scanners typically capture data by successively considering points on a discrete, regular (typically spherical) raster, and recording the respective geometric and radiometric information. In contrast, range cameras have an internal sensor array for simultaneously considering all points on a discrete, regular (typically planar) raster, and recording the respective geometric and radiometric information. If the data acquisition is performed sufficiently fast, range cameras also allow an acquisition of dynamic scenes. For both types of acquisition systems, however, the sampling of data on a discrete, regular raster allows to derive 2D image representations for spatial 3D information and the respective radiometric information. While the spatial 3D information and thus its 2D representation in the form of a range image, where each pixel represents the distance r between the scanning device and the respective 3D scene point, depend on the scene structure, the captured radiometric information depends on both the scene structure and the signal processing unit in the involved acquisition system. Note that differences with respect to the signal processing unit mainly arise from internal processes such as the conversion to a digital signal and signal amplification which are not identical for different scanning devices.

In order to derive a representation which is independent of the involved scanning device, it is feasible to transform the radiometric information representing the measured energy of the backscattered laser light from arbitrary values to a given interval of possible values. In this context, the radiometric information represented in a matrix $\mathbf{I} \in \mathbb{R}^{n \times m}$ with n rows and m columns is typically transformed to an 8-bit gray-valued image $\mathbf{I}_{\text{norm}}$ by performing a histogram normalization of the type

$$\mathbf{I}_{\text{norm}} = \text{uint8} \left(255 \cdot \frac{\mathbf{I} - I_{\min} \cdot \mathbf{1}_{n \times m}}{I_{\max} - I_{\min}} \right) \qquad (2.1)$$

where $I_{\min}, I_{\max} \in \mathbb{R}$ are the minimum and maximum values of $\mathbf{I}$, respectively. Thus, the elements of the resulting $(n \times m)$-matrix $\mathbf{I}_{\text{norm}}$ are represented by integer values in the interval $[0, 255]$.

Considering the adapted radiometric information, we may describe a scan $\mathscr{S} = \{\mathscr{I}_{\text{R}}, \mathscr{I}_{\text{I}}\}$ with respective 2D representations in the form of a range image $\mathscr{I}_{\text{R}} \in \mathbb{R}^{n \times m}$ and an intensity image $\mathscr{I}_{\text{I}} = \mathbf{I}_{\text{norm}} \in \mathbb{R}^{n \times m}$. On both types of images, we may apply standard image processing techniques, e.g., to detect specific features or to recognize and segment objects. In order to obtain an impression about the characteristics of such images, we consider a scan taken from a benchmark dataset[4] [9] which is later

---

[4]This dataset is available at http://www.ikg.uni-hannover.de/index.php?id=413\&L=de (last access: 30 May 2015).

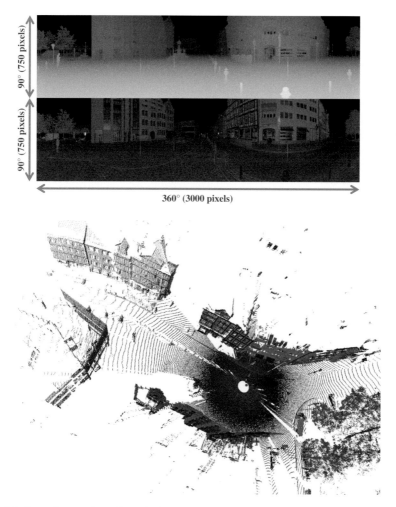

Fig. 2.3 Range image $\mathscr{I}_R$ and intensity image $\mathscr{I}_I$ (*top*), and a visualization of the scan as 3D point cloud where the position of the scanning device is indicated by a *red dot* (*bottom*)

used for evaluating novel approaches for point cloud registration in Chap. 4. The respective range image $\mathscr{I}_R$ as well as the respective intensity image $\mathscr{I}_I$ are shown in Fig. 2.3.

2.5 Point Quality Assessment

When dealing with 3D point cloud data, it should be taken into account that, usually, a filtering of the raw point cloud data is carried out by internal software provided by the manufacturer of a device. Thus, the applied methodology is hidden in a "black box" whose procedures may hardly be reconstructed. Consequently, we aim to found our methodology on raw 3D point cloud data and, hence, understand the routines applied for filtering 3D point cloud data by removing unreliable range measurements. In order to do so, we first reflect different factors which influence the quality of a range measurement as well as existing approaches for quantifying the quality of a range measurement (Sect. 2.5.1). Based on these considerations, we proceed with describing a standard approach based on intensity information (Sect. 2.5.2) and, subsequently, we present two novel approaches based on *range reliability* (Sect. 2.5.3) and *local planarity* (Sect. 2.5.4) which have been developed in the scope of our work. In order to reason about their suitability, we provide a qualitative comparison (Sect. 2.5.5) as well as a quantitative comparison (Sect. 2.5.6) of these approaches.

2.5.1   Influencing Factors and Related Work

Generally, the quality of a range measurement depends on a variety of influencing factors. Following seminal work on categorizing these factors [6, 18, 19, 43], we take into account four different categories:

- The first category of influencing factors is based on the design of the acquisition system which, according to [21, 44], may be characterized by the laser type, the modulation technique (e.g., continuous wave or pulsed lasers), the measurement principle (e.g., time-of-flight, phase or amplitude measurements), the detection technique (e.g., coherent or direct detection), and the configuration between transmitting and receiving unit (e.g., monostatic or bistatic configuration). Concerning these components, specific factors influencing individual point quality mainly arise from respective hardware properties (e.g., angular resolution, angular accuracy, and range accuracy) and the calibration of the used device (e.g., calibration model or short-term/long-term stability).
- The second category of influencing factors addresses atmospheric and environmental conditions [7, 49]. While uncertainties due to atmospheric conditions mainly arise from humidity, temperature, aerosols (i.e., fine solid particles or liquid droplets in the air), or variations in air pressure, the environmental conditions address the presence of ambient lighting (i.e., natural sunlight or artificial light) and the scene type (i.e., indoor or outdoor scene).
- The third category of influencing factors addresses characteristics of the observed scene in terms of object materials, surface reflectivity, and surface roughness which strongly influence light reflection on the object surfaces [27, 48, 57].
- The fourth category of influencing factors addresses the scanning geometry [2, 35], i.e., the distance and orientation of scanned surfaces with respect to the involved sensor. Particularly the incidence angle representing the angle between incoming laser beam and surface normal strongly influences individual point quality [42, 43], but also those effects occurring at object edges have to be addressed [4].

While a systematic error modeling may account for potential error sources arising from the acquisition system [3, 8, 28, 29], scene-specific issues cannot be generalized and therefore have to be treated differently. In the scope of our work, we focus on terrestrial laser scanning and mobile laser scanning in outdoor environments with reasonable natural sunlight. Consequently, we face relatively small distances between scene objects and the acquisition system as well as relatively small displacements of the acquisition system for obtaining consecutive scans, so that atmospheric and environmental conditions may be neglected. Furthermore, we do not make assumptions on the presence of specific object materials and the corresponding surface reflectivity, since we focus on digitizing arbitrary scenes without involving prior knowledge about the presence and the surface reflectance behavior of specific materials. As a consequence, we mainly focus on the scanning geometry and the reflected energy (represented as intensity value) per range measurement. Thus, we may treat the quantification of the quality of a range measurement decoupled in a post-processing step.

The decoupled consideration may also directly be applied for available benchmark datasets, where we may purely focus on filtering available raw 3D point cloud data by exploiting the captured information. On the one hand, a simple approach for removing unreliable range measurements may be based on the captured intensity information [4], since very low intensity values are likely to correspond to unreliable range measurements. On the other hand, it seems desirable to explicitly address points at depth discontinuities as these exhibit the largest uncertainty in distance. A respective filtering may, for instance, be achieved by considering the variance of the local normal vector [2], by removing 3D points corresponding to nonplanar objects or objects which are susceptible to occlusions, shadows, etc. [16, 17], by applying the Laplacian operator on the range image [5] or by involving the scan line approximation technique [14]. Accounting for both intensity and range information, the combination of removing points with low values in the intensity image as well as points at edges in the range image has been proposed in order to obtain an adequate 3D representation of a scene [45]. Furthermore, it should be taken into account that the scanning geometry with respect to the incidence angle may have a significant influence on the accuracy of a range measurement which becomes visible by an increase in measurement noise with increasing incidence angles [43]. Accordingly, it would be desirable to have a common and generic measure which considers reliability in terms of both object edges and incidence angle.

2.5.2   Filtering Based on Intensity Information

As already mentioned above, a filtering of raw 3D point cloud data may be based on intensity information [4], since low intensity values typically indicate unreliable range measurements. Consequently, the main challenge of such an approach consists of finding a suitable threshold in order to distinguish reliable range measurements from unreliable ones. Such a threshold may for instance be derived by applying a histogram-based approach estimating a suitable border between bimodal distributions. However, the measured intensity values may be arbitrary and significantly vary from one device to another, since they depend on the electronic circuits inside the scanning device. A transfer toward a general application across a variety of different scanning devices may easily be derived by applying a histogram normalization and thereby mapping the measured intensity values onto a predefined fixed interval whose borders may principally be selected arbitrarily. Accordingly, we may for instance select the interval [0, 1], but—as we have already derived a 2D image representation of intensity information in the form of an 8-bit gray-valued image (Sect. 2.4)—it is feasible to use the interval [0, 255] instead and select the threshold as a specific gray value $t_{I,gray}$. Note that such a simple thresholding based on intensity information may also be used to detect and remove soft obstacles which are for instance given by smoke, dust, fog or steam [11].

While an intensity-based filtering of raw 3D point cloud data is straightforward, easy-to-implement, and quite efficient,[5] such considerations do not account for edge effects where noisy range measurements are likely to occur although the respective intensity values might be reasonable. Hence, we focus on two novel strategies which are based on the geometric measures of (i) *range reliability* and (ii) *local planarity* in order to quantify the quality of a range measurement.

2.5.3   Filtering Based on Range Reliability

The first of our proposed measures to quantify the reliability of a range measurement directly addresses the fact that—instead of considering a laser beam as a ray with no physical dimensions—it is inevitable that a laser beam has certain physical dimensions which, in turn, influences the laser beam propagation. Thus, the projection of a laser beam on the target area results in a *laser footprint*, i.e., a spot with finite dimension, that may vary depending on the slope of the local surface and material characteristics [50]. Consequently, if a measured 3D point corresponds to a laser footprint on a geometrically smooth surface, the captured range information is rather reliable when assuming Lambertian surfaces and reasonable incidence angles. However, at edges of objects, a laser footprint may cover surfaces at different distances to the sensor, and thus the captured range information is rather unreliable. Even more critical are range measurements corresponding to the sky, since these mainly arise from atmospheric effects.

In order to quantify these effects and remove unreliable range measurements— which would typically appear as noisy behavior in a visualization of the considered 3D point cloud data—our first measure referred to as *range reliability* [51] is based on the categorization of each point on the 2D scan grid by considering local image

---

[5]For our image-based consideration, we only need to calculate the sign-function of the difference of the intensity image $\mathscr{I}_I$ and an image $\mathscr{I}_I^* = t_{I,gray} \cdot \mathbf{1}_{n \times m} \in \mathbb{R}^{n \times m}$ of constant gray value $t_{I,gray}$, which yields pixel-wise conclusions about reliable or unreliable range measurements.

| 0 | 15 | 0 | 255 | 0.01 | 40.00 | log(0.01) | log(40.00) | unreliable reliable |

Fig. 2.4 Range image $\mathscr{I}_R$, intensity image $\mathscr{I}_I$, visualization of range reliability $\sigma_{r,3\times3}$, logarithmic representation of range reliability $\sigma_{r,3\times3}$ and the binary confidence map derived by thresholding based on a manually selected threshold of $t_\sigma = 0.03$ m (from *left* to *right*)

patches in the respective range image. More specifically, this measure considers a local (3×3) image neighborhood for each pixel (x, y) in the range image and assigns the standard deviation $\sigma_{r,3\times3}$ of all range values r within the (3×3) image neighborhood to (x, y). Deriving $\sigma_{r,3\times3}$ for all pixels of the range image yields a confidence map. In this confidence map, low values $\sigma_{r,3\times3}$ indicate a 3D point on a smooth surface and are therefore assumed to correspond to reliable range measurements, whereas high values $\sigma_{r,3\times3}$ indicate noisy and unreliable range measurements. Consequently, a simple thresholding based on a predefined threshold t_σ may be considered as sufficient to separate reliable measurements from unreliable ones in a binary confidence map, which is shown in Fig. 2.4 for a part of a terrestrial laser scan which corresponds to 2304×1135 scanned 3D points and has been acquired with a Leica HDS6000 on the KIT campus in Karlsruhe, Germany. According to qualitative tests involving different scanning devices, a value of $t_\sigma = 0.03 \ldots 0.10$ m has proven to be appropriate for data captured with a terrestrial laser scanner [51] and for data captured with a range camera [54, 55], but the manual selection of a threshold based on prior knowledge about the scene and/or data represents a limitation.

2.5.4 Filtering Based on Local Planarity

The second of our proposed measures which we refer to as *local planarity* [52] is motivated by the fact that reliable range information typically corresponds to almost planar structures in the scene which are characterized by low incidence angles. Consequently, we aim to quantify the local planarity for each point on the 2D scan grid by considering local image patches in the respective range image. In analogy to the measure of range reliability, we consider (3×3) image neighborhoods as local image patches in order to assign a measure of planarity to the respective center point. From the spatial *XYZ*-coordinates of all 3D points corresponding to the pixels in the (3×3) image neighborhood, we derive the 3D covariance matrix known as *3D structure tensor* $\mathbf{S}_{3D} \in \mathbb{R}^{3\times 3}$ whose eigenvalues $\lambda_1, \lambda_2, \lambda_3 \in \mathbb{R}$ with $\lambda_1 \geq \lambda_2 \geq \lambda_3 \geq 0$ are further exploited in order to define the *dimensionality features* of *linearity* L_λ, *planarity* P_λ and *scattering* S_λ [58]:

$$L_\lambda = \frac{\lambda_1 - \lambda_2}{\lambda_1} \qquad P_\lambda = \frac{\lambda_2 - \lambda_3}{\lambda_1} \qquad S_\lambda = \frac{\lambda_3}{\lambda_1} \qquad (2.2)$$

These dimensionality features reveal a normalization by the largest eigenvalue λ_1, so that they sum up to 1 and the largest value among the dimensionality features indicates the characteristic behavior of the respective pixel. Accordingly, a pixel (x, y) in the range image represents a planar 3D structure and therefore rather reliable range information if the local planarity $P_{\lambda,3\times 3}$ in a (3×3) image neighborhood satisfies the constraint

$$P_{\lambda,3\times 3} := P_\lambda \geq \max \{L_\lambda, S_\lambda\} \qquad (2.3)$$

which—in contrast to the measure of range reliability—yields a binary confidence map (e.g., as illustrated in Fig. 2.5) in a fully generic manner without involving any manually specified thresholds and thus prior knowledge about the scene and/or data.

Note that the dimensionality features could also directly be extracted for each point of the 3D point cloud by considering other 3D points in the local 3D neighborhood and exploiting the respective 3D structure tensor in order to derive the eigenvalues. However, particularly for large point clouds, it can be quite time-consuming to extract suitable neighborhoods and, in turn, the dimensionality features resulting from the eigenvalues of the 3D structure tensor. Consequently, the proposed consideration of

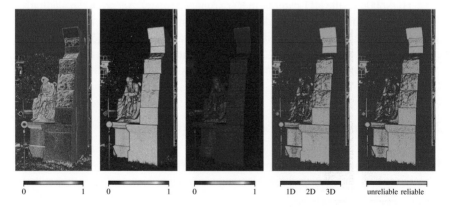

| 0 | 1 | 0 | 1 | 0 | 1 | 1D | 2D | 3D | unreliable | reliable |

Fig. 2.5 Visualization of linearity L_λ, planarity P_λ, scattering S_λ, the classification of each pixel according to its local behavior (linear (1D): *red*; planar (2D): *green*; scattered (3D): *blue*) and the derived binary confidence map (from *left* to *right*)

local planarity $P_{\lambda,3\times3}$ based on a local (3×3) image neighborhood offers the great advantage that the direct neighbors of each point on the regular 2D scan grid may be derived more efficiently with respect to processing time and memory consumption than a suitable number of neighbors in 3D.

2.5.5 A Qualitative Comparison of Different Measures

In order to provide an impression about the performance of the different measures for quantifying the quality of range measurements, we consider 2D image representations for range and intensity information in Fig. 2.6 as well as the respective binary confidence maps based on (i) intensity information when applying a threshold of $t_{I,gray} = 10$ (for an 8-bit gray-valued image), (ii) the measure $\sigma_{r,3\times3}$ of range reliability when applying a threshold of $t_\sigma = 0.03$ m, and (iii) the generic measure $P_{\lambda,3\times3}$ of local planarity. For each measure, the corresponding effect in 3D space is visualized in Fig. 2.7. This figure clearly reveals that the use of intensity information alone is not sufficient to adequately filter raw 3D point cloud data and thereby completely remove the noisy behavior. In contrast, the strategies based on the two geometric measures retain adequate representations of local object surfaces. Whereas the strategy based on the measure $\sigma_{r,3\times3}$ of range reliability provides almost planar object surfaces for significantly varying incidence angles (Fig. 2.6), the strategy based on the measure $P_{\lambda,3\times3}$ of local planarity only provides almost perpendicular object surfaces with almost planar behavior and thus favors lower incidence angles which tend to yield more accurate range measurements. Note that, even though the sphere target in the lower left part of the depicted images does not provide a planar surface, the measures of range reliability and local planarity indicate reliable range measurements for almost all points corresponding to the surface of the sphere target, and only points

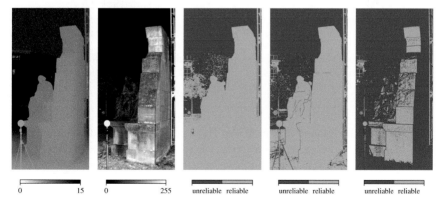

Fig. 2.6 Range image $\mathscr{I}_R$, intensity image $\mathscr{I}_I$, and the derived binary confidence maps based on intensity information, range reliability $\sigma_{r,3\times3}$ and local planarity $P_{\lambda,3\times3}$ (from *left* to *right*)

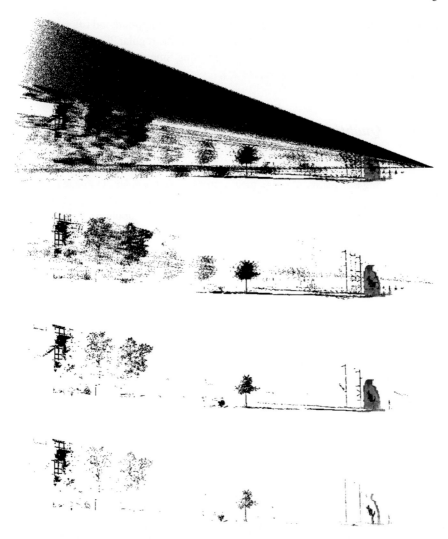

Fig. 2.7 Raw 3D point cloud data, 3D point cloud data filtered via intensity information, 3D point cloud data filtered via the measure $\sigma_{r,3\times3}$ of range reliability and 3D point cloud data filtered via the measure $P_{\lambda,3\times3}$ of local planarity (from *top* to *bottom*)

at the edges are considered as unreliable. This is due to the consideration of relatively small (3×3) image neighborhoods, where no significant deviation in range direction and no significant deviation from a locally planar surface may be observed.

2.5.6 A Quantitative Comparison of Different Measures

In this section, we intend to quantitatively compare the different measures for point quality assessment and verify our observations made in the previous subsection. For this purpose, we consider a fairly simple scenario and carry out theoretical considerations for the proposed measures of range reliability $\sigma_{r,3\times3}$ and local planarity $P_{\lambda,3\times3}$ and, thereby, we aim to point out consequences concerning what we may expect when applying these measures on range images. This is of utmost importance since we may thus easily explain the significant differences between the binary confidence maps depicted in Figs. 2.4 and 2.5.

The considered scenario focuses on the characteristics of scanned 3D points on a planar surface with a certain distance and a certain incidence angle with respect to the scanning device. Considering fundamentals of projective geometry as for instance described in [15], the 3D coordinates of any point $\mathbf{X} \in \mathbb{R}^3$ on a ray in 3D space satisfy the constraint $\mathbf{X} = \mathbf{A} + b\mathbf{v}$, where $\mathbf{A} \in \mathbb{R}^3$ denotes a known 3D point on the ray, $b \in \mathbb{R}$ represents a scalar factor and $\mathbf{v} \in \mathbb{R}^3$ indicates the direction of the ray. Without loss of generality, we may transfer this equation from world coordinates to camera coordinates as indicated by a superscript ^{c}, i.e., $\mathbf{X}^c = \mathbf{A}^c + b\mathbf{v}^c$. Note that assuming the model of a pinhole camera is valid when using range cameras and approximately valid when considering a local (3×3) neighborhood on a spherical scan grid with reasonable angular resolution. Since the considered rays thus intersect each other at the projective center $\mathbf{0}^c$, it is straightforward to use the point $\mathbf{A}^c = \mathbf{0}^c = [0, 0, 0]^T$ as known 3D point on all rays. Following the standard definitions, we may furthermore define the camera coordinate frame relative to the device in a way so that the X^c-axis points to the right, the Y^c-axis to the bottom and the Z^c-axis in depth. Considering a local (3×3) image neighborhood, we may conduct point quality assessment by looking along the Z^c-axis and assuming an angular resolution $\Delta\alpha$ of the camera. Accordingly, the directions $\mathbf{v}^c$ of the 8 neighboring rays which are exploited to obtain a local (3×3) image neighborhood may easily be derived by an intersection with the $(Z^c = 1)$-plane and we thus evaluate the geometric behavior of range measurements in a field-of-view given by $(2\Delta\alpha \times 2\Delta\alpha)$. Note that, in this context, typical angular resolutions would be $\Delta\alpha \approx \{0.36°$ (middle scan density), $0.18°$ (high scan density), $0.09°$ (super high scan density), $0.05°$ (ultra high scan density)$\}$ for a terrestrial laser scanner of type Leica HDS6000, $\Delta\alpha \approx 0.2°$ for a range camera of type PMD[vision] CamCube 2.0 and $\Delta\alpha \approx 0.09°$ for a range camera of type Microsoft Kinect.

Once the rays have been derived for a field-of-view of $(2\Delta\alpha \times 2\Delta\alpha)$, we may proceed by assuming that these rays intersect a plane π which is parameterized in the camera coordinate frame by a 3D point $\mathbf{X}^c_\pi$ and a normal vector $\mathbf{n}^c_\pi \in \mathbb{R}^3$. Thereby, we define the 3D point $\mathbf{X}^c_\pi$ as the point which results from the intersection of π with the Z^c-axis, and we further assume that the distance between $\mathbf{X}^c_\pi$ and $\mathbf{0}^c$ is given by d, i.e., $\mathbf{X}^c_\pi = [0, 0, d]^T$. Initially, we consider the case of a normal vector $\mathbf{n}^c_\pi$ which coincides with the Z^c-axis, and thus the plane π is perpendicular to the Z^c-axis and parallel to the X^cY^c-plane. Subsequently, we rotate the plane π by an angle β around the axis defined by the point $\mathbf{X}^c_\pi = [0, 0, d]^T$ and the direction $[0, 1, 0]^T$ in

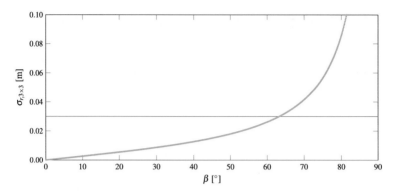

Fig. 2.8 Behavior of the measure $\sigma_{r,3\times3}$ of range reliability for increasing incidence angles β. The applied threshold of $t_\sigma = 0.03$ m is indicated with a *red line*

the camera coordinate frame. As a result, the angle β coincides with the incidence angle and we get 9 points of intersection for angles $\beta \in [0°, 90°)$.

In order to verify the suitability of the proposed measure $\sigma_{r,3\times3}$ of range reliability, we simply exploit the distances between the projective center $\mathbf{0}^c$ and the 9 points of intersection. These distances correspond to range measurements and the measure $\sigma_{r,3\times3}$ of range reliability (see Sect. 2.5.3) simply represents the standard deviation of the 9 considered distances (i.e., range values). For a representative example, we consider the use of a range camera of type PMD[vision] CamCube 2.0 and, accordingly, we select the angular resolution to $\Delta\alpha = 0.2°$ and the distance between projective center $\mathbf{0}^c$ and $\mathbf{X}_\pi^c$ to $d = 5$ m. The respective behavior of $\sigma_{r,3\times3}$ for varying incidence angles is visualized in Fig. 2.8. This figure clearly reveals that, when applying the proposed threshold of $t_\sigma = 0.03$ m, range measurements are assumed to be reliable for incidence angles of less than about 63.3°. A less strict threshold of $t_\sigma = 0.10$ m even results in reliable range measurements up to incidence angles of about 81.4°.

In order to verify the suitability of the proposed measure $P_{\lambda,3\times3}$ of local planarity, we exploit the 3D coordinates of the 9 points of intersection in order to derive the 3D structure tensor $\mathbf{S}_{3D}$ and its eigenvalues λ_1, λ_2 and λ_3 as well as the dimensionality features of linearity L_λ, planarity P_λ and scattering S_λ (see Sect. 2.5.4). For a representative example, we again consider the use of a range camera of type PMD[vision] CamCube 2.0 and, accordingly, we select the angular resolution to $\Delta\alpha = 0.2°$ and the distance between projective center $\mathbf{0}^c$ and $\mathbf{X}_\pi^c$ to $d = 5$ m. The respective values of the dimensionality features for angles $\beta \in [0°, 90°)$ are depicted in Fig. 2.9, and they reveal that the locally planar 3D structure provides a planar behavior in the interval $[0°, 45°]$ and a linear behavior beyond this interval. As a consequence, range measurements are assumed to be reliable if the local (3×3) image neighborhood represents a locally planar 3D structure with an incidence angle in $[0°, 45°]$. Note that, due to the narrow field-of-view of $(2\Delta\alpha \times 2\Delta\alpha)$ for a local (3×3) image patch, noisy range measurements e.g., corresponding to the sky will not be indicated by a scattered behavior, but by a linear behavior since only a significant variation in ray

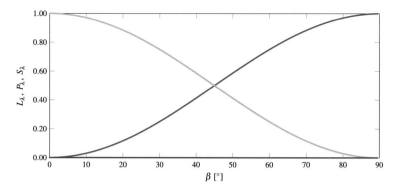

Fig. 2.9 Behavior of the dimensionality features of linearity L_λ (*red*), planarity P_λ (*green*), and scattering S_λ (*blue*) for increasing incidence angles β

direction will be present. Furthermore considering that the dimensionality features are normalized by the largest eigenvalue, our observations on reliable range measurements in terms of local planarity are independent from both the angular resolution $\Delta\alpha$ and the distance between $\mathbf{X}_\pi^c$ and $\mathbf{0}^c$.

Based on these results, the considered scenario allows to provide qualitative statements on the suitability of both measures. The binary confidence map shown in Fig. 2.4 provides more reliable range measurements than the binary confidence map depicted in Fig. 2.5, particularly for those scanned 3D points on planar surfaces with a higher incidence angle. The reason for this behavior becomes visible when comparing Figs. 2.8 and 2.9, since applying the measure $\sigma_{r,3\times3}$ of range reliability also allows incidence angles which are much larger than 45°, whereas the measure $P_{\lambda,3\times3}$ of local planarity relates reliable range measurements to planar surfaces with incidence angles of up to only about 45°. Furthermore, it should be taken into account that, while the measure $\sigma_{r,3\times3}$ of range reliability depends on the distances between the projective center and the 9 points of intersection as well as the angular resolution $\Delta\alpha$ of the scanning device, the measure $P_{\lambda,3\times3}$ of local planarity is independent from both of them.

2.6 Conclusions

In this chapter, we have described fundamentals concerning the definition of 3D point clouds and their acquisition. Focusing on active point cloud acquisition via laser scanners and range cameras, we have addressed the issue of how to transform the captured 3D point cloud data to 2D representations in the form of range images and intensity images, since such images allow to exploit a rich variety of approaches which have proven to be effective and efficient for tasks such as feature extraction, object recognition, or object segmentation. Furthermore, we have provided an overview on techniques for judging about the quality of each 3D point by considering the reliability of the respective range measurement and, in this context, we have presented two novel approaches and discussed their capabilities. From a comparison of 3D visualizations of raw 3D point cloud data and, respectively, filtered 3D point cloud data for an exemplary TLS scan, we may conclude that a filtering based on point quality assessment might not only visually improve the results, but also alleviate subsequent tasks relying on either all or only some of the acquired 3D points. Later, in Chap. 4, we will make use of the presented concepts for (i) generating 2D representations and (ii) assessing point quality in the context of point cloud registration, and we will see significant advantages arising from such concepts.

References

1. Agarwal S, Snavely N, Simon I, Seitz SM, Szeliski R (2009) Building Rome in a day. In: Proceedings of the IEEE international conference on computer vision, pp 72–79
2. Bae K-H, Belton D, Lichti DD (2005) A framework for position uncertainty of unorganised three-dimensional point clouds from near-monostatic laser scanners using covariance analysis. Int Arch Photogramm Remote Sens Spat Inf Sci XXXVI-3/W19:7–12
3. Barber D, Mills J, Smith-Voysey S (2008) Geometric validation of a ground-based mobile laser scanning system. ISPRS J Photogramm Remote Sens 63(1):128–141
4. Barnea S, Filin S (2007) Registration of terrestrial laser scans via image based features. Int Arch Photogramm Remote Sens Spat Inf Sci XXXVI-3/W52:32–37
5. Barnea S, Filin S (2008) Keypoint based autonomous registration of terrestrial laser point-clouds. ISPRS J Photogramm Remote Sens 63(1):19–35
6. Boehler W, Bordas Vicent M, Marbs A (2003) Investigating laser scanner accuracy. Int Arch Photogramm Remote Sens Spat Inf Sci XXXIV-5/C15:1–6
7. Borah DK, Voelz DG (2007) Estimation of laser beam pointing parameters in the presence of atmospheric turbulence. Appl Opt 46(23):6010–6018
8. Boström G, Gonçalves JGM, Sequeira V (2008) Controlled 3D data fusion using error-bounds. ISPRS J Photogramm Remote Sens 63(1):55–67
9. Brenner C, Dold C, Ripperda N (2008) Coarse orientation of terrestrial laser scans in urban environments. ISPRS J Photogramm Remote Sens 63(1):4–18
10. Dal Mutto C, Zanuttigh P, Cortelazzo GM (2012) Time-of-flight cameras and Microsoft Kinect(TM). Springer, New York
11. Djuricic A, Jutzi B (2013) Supporting UAVs in low visibility conditions by multiple-pulse laser scanning devices. Int Arch Photogramm Remote Sens Spat Inf Sci XL-1/W1:93–98
12. Droeschel D, Holz D, Behnke S (2010) Multi-frequency phase unwrapping for time-of-flight cameras. In: Proceedings of the IEEE/RSJ international conference on intelligent robots and systems, pp 1463–1469
13. Frahm J-M, Fite-Georgel P, Gallup D, Johnson T, Raguram R, Wu C, Jen Y-H, Dunn E, Clipp B, Lazebnik S, Pollefeys M (2010) Building Rome on a cloudless day. In: Proceedings of the European conference on computer vision, vol IV, pp 368–381
14. Fuchs S, May S (2008) Calibration and registration for precise surface reconstruction with time-of-flight cameras. Int J Intell Syst Technol Appl 5(3/4):274–284
15. Hartley RI, Zisserman A (2008) Multiple view geometry in computer vision. University Press, Cambridge
16. Hebel M, Stilla U (2007) Automatic registration of laser point clouds of urban areas. Int Arch Photogramm Remote Sens Spat Inf Sci XXXVI-3/W49A:13–18

17. Hebel M, Stilla U (2009) Automatische Koregistrierung von ALS-Daten aus mehreren Schräg-ansichten städtischer Quartiere. PFG—Photogramm Fernerkund Geoinf 3(2009):261–275
18. Hebert M, Krotkov E (1992) 3-D measurements from imaging laser radars: how good are they? Int J Image Vis Comput 10(3):170–178
19. Hennes M (2007) Konkurrierende Genauigkeitsmaße—Potential und Schwächen aus der Sicht des Anwenders. AVN—Allg Vermess-Nachr 4(2007):136–146
20. Herbort S, Wöhler C (2011) An introduction to image-based 3D surface reconstruction and a survey of photometric stereo methods. 3D Res 2(3):40:1–40:17
21. Jutzi B (2007) Analyse der zeitlichen Signalform von rückgestreuten Laserpulsen. PhD thesis, Institut für Photogrammetrie und Kartographie, Technische Universität München, München, Germany
22. Jutzi B (2009) Investigations on ambiguity unwrapping of range images. Int Arch Photogramm Remote Sens Spat Inf Sci XXXVIII-3/W8:265–270
23. Jutzi B (2012) Extending the range measurement capabilities of modulated range imaging devices by time-frequency-multiplexing. AVN—Allg Vermess-Nachr 2(2012):54–62
24. Jutzi B (2015) Methoden zur automatischen Szenencharakterisierung basierend auf aktiven optischen Sensoren für die Photogrammetrie und Fernerkundung. Habilitation thesis, Institute of Photogrammetry and Remote Sensing, Karlsruhe Institute of Technology (KIT), Karlsruhe, Germany
25. Khoshelham K, Oude Elberink S (2012) Accuracy and resolution of Kinect depth data for indoor mapping applications. Sensors 12(2):1437–1454
26. Kolb A, Barth E, Koch R, Larsen R (2010) Time-of-flight cameras in computer graphics. Comput Graph Forum 29(1):141–159
27. Leader JC (1979) Analysis and prediction of laser scattering from rough-surface materials. J Opt Soc Am A 69(4):610–628
28. Lichti DD, Gordon SJ, Tipdecho T (2005) Error models and propagation in directly georefer-enced terrestrial laser scanner networks. J Survey Eng 131(4):135–142
29. Lichti DD, Licht MG (2006) Experiences with terrestrial laser scanner modelling and accuracy assessment. Int Arch Photogramm Remote Sens Spat Inf Sci XXXVI-5:155–160
30. Moons T, Van Gool L, Vergauwen M (2010) 3D reconstruction from multiple images—Part 1: principles. Found Trends Comput Graph Vis 4(4):287–404
31. Remondino F, Stoppa D (2013) TOF range-imaging cameras. Springer, Heidelberg
32. Rusu RB, Cousins S (2011) 3D is here: Point Cloud Library (PCL). In: Proceedings of the IEEE international conference on robotics and automation, pp 1–4
33. Salvi J, Pagès J, Batlle J (2004) Pattern codification strategies in structured light systems. Pattern Recognit 37(4):827–849
34. Sansoni G, Trebeschi M, Docchio F (2009) State-of-the-art and applications of 3D imaging sensors in industry, cultural heritage, medicine, and criminal investigation. Sensors 9(1):568–601
35. Schaer P, Legat K, Landtwing S, Skaloud J (2007) Accuracy estimation for laser point cloud including scanning geometry. Int Arch Photogramm Remote Sens Spat Inf Sci XXXVI-5/C55:1–8
36. Scharstein D, Szeliski R (2002) A taxonomy and evaluation of dense two-frame stereo corre-spondence algorithms. Int J Comput Vis 47(1–3):7–42
37. Seitz SM, Curless B, Diebel J, Scharstein D, Szeliski R (2006) A comparison and evaluation of multi-view stereo reconstruction algorithms. In: Proceedings of the IEEE computer society conference on computer vision and pattern recognition, vol 1, pp 519–528
38. Smisek J, Jancosek M, Pajdla T (2011) 3D with Kinect. In: Proceedings of the IEEE interna-tional conference on computer vision workshops, pp 1154–1160
39. Soatto S (2009) Actionable information in vision. In: Proceedings of the IEEE international conference on computer vision, pp 2138–2145
40. Soatto S (2011) Steps towards a theory of visual information: active perception, signal-to-symbol conversion and the interplay between sensing and control. Technical Report, arXiv:1110.2053 [cs.CV]

41. Soudarissanane S, Lindenbergh R (2011) Optimizing terrestrial laser scanning measurement set-up. Int Arch Photogramm Remote Sens Spat Inf Sci XXXVIII-5/W12:127–132
42. Soudarissanane S, Lindenbergh R, Menenti M, Teunissen P (2009) Incidence angle influence on the quality of terrestrial laser scanning points. Int Arch Photogramm Remote Sens Spat Inf Sci XXXVIII-3/W8:183–188
43. Soudarissanane S, Lindenbergh R, Menenti M, Teunissen P (2011) Scanning geometry: influencing factor on the quality of terrestrial laser scanning points. ISPRS J Photogramm Remote Sens 66(4):389–399
44. Stilla U, Jutzi B (2009) Waveform analysis for small-footprint pulsed laser systems. In: Shan J, Toth CK (eds) Topographic laser ranging and scanning: principles and processing. CRC Press, Boca Raton, pp 215–234
45. Swadzba A, Liu B, Penne J, Jesorsky O, Kompe R (2007) A comprehensive system for 3D modeling from range images acquired from a 3D ToF sensor. In: Proceedings of the international conference on computer vision systems, pp 1–10
46. Szeliski R (2011) Computer vision: algorithms and applications. Springer, London
47. Triggs B, McLauchlan PF, Hartley RI, Fitzgibbon AW (1999) Bundle adjustment—A modern synthesis. In: Proceedings of the international workshop on vision algorithms: theory and practice, pp 298–372
48. Voegtle T, Schwab I, Landes T (2008) Influences of different materials on the measurements of a terrestrial laser scanner (TLS). Int Arch Photogramm Remote Sens Spat Inf Sci XXXVII-B5:1061–1066
49. Voisin S, Foufou S, Truchetet F, Page D, Abidi M (2007) Study of ambient light influence for three-dimensional scanners based on structured light. Opt Eng 46(3):030502:1–3
50. Vosselman G, Maas H-G (2010) Airborne and terrestrial laser scanning. Whittles Publishing, Dunbeath
51. Weinmann M, Jutzi B (2011) Fully automatic image-based registration of unorganized TLS data. Int Arch Photogramm Remote Sens Spat Inf Sci XXXVIII-5/W12:55–60
52. Weinmann M, Jutzi B (2015) Geometric point quality assessment for the automated, markerless and robust registration of unordered TLS point clouds. ISPRS Ann Photogramm Remote Sens Spat Inf Sci II-3/W5:89–96
53. Weinmann M, Wursthorn S, Jutzi B (2011) Semi-automatic image-based fusion of range imaging data with different characteristics. Int Arch Photogramm Remote Sens Spat Inf Sci XXXVIII-3/W22:119–124
54. Weinmann M, Dittrich A, Hinz S, Jutzi B (2013) Automatic feature-based point cloud registration for a moving sensor platform. Int Arch Photogramm Remote Sens Spat Inf Sci XL-1/W1:373–378
55. Weinmann M, Leitloff J, Hoegner L, Jutzi B, Stilla U, Hinz S (2014) Thermal 3D mapping for object detection in dynamic scenes. ISPRS Ann Photogramm Remote Sens Spat Inf Sci II-1:53–60
56. Weinmann Mi, Osep A, Ruiters R, Klein R (2013) Multi-view normal field integration for 3D reconstruction of mirroring objects. In: Proceedings of the IEEE international conference on computer vision, pp 2504–2511
57. Weinmann Mi, Klein R (2015) Advances in geometry and reflectance acquisition (course notes). In: Proceedings of the SIGGRAPH Asia 2015 Courses, pp 1:1–1:71
58. West KF, Webb BN, Lersch JR, Pothier S, Triscari JM, Iverson AE (2004) Context-driven automated target detection in 3-D data. Proc SPIE 5426:133–143
59. Wursthorn S (2014) Kamerabasierte Egomotion-Bestimmung mit natürlichen Merkmalen zur Unterstützung von Augmented-Reality-Systemen. PhD thesis, Institute of Photogrammetry and Remote Sensing, Karlsruhe Institute of Technology (KIT), Karlsruhe, Germany

Chapter 3
A Brief Survey on 2D and 3D Feature Extraction

The detection, description, and comparison of data contents is of great importance for a variety of research domains. In this chapter, we intend to enlighten these issues in the context of 2D imagery and 3D point cloud data. Hence, we address fundamental ideas by focusing on several questions: How can we detect an object? How can we describe an object? What makes an object memorable? According to which criteria can we recognize the same object or similar objects? How could similarity be defined? In order to answer such questions which, in turn, allow us to infer a deeper understanding of the respective scene, we describe how distinctive characteristics contained in respective 2D or 3D data may be represented as *features*, and we thereby categorize features according to different feature types [41, 42]. For this purpose, we first derive a general definition for characterizing a feature (Sect. 3.1). Subsequently, we focus on feature extraction from 2D imagery (Sect. 3.2) as well as feature extraction from 3D point cloud data (Sect. 3.3). Based on these findings, we discuss the motivation for involving specific features in the scope of our work (Sect. 3.4) and, finally, we provide concluding remarks (Sect. 3.5).

3.1 What Is a Feature?

As a first step, we focus on deriving general characteristics of a feature. Following our basic concept presented in Sect. 2.1, we may generally consider a *feature* as a specific representation of data. In this regard, a feature typically consists of a compact representation of global or local properties of the given data, and the suitability of a specific feature may strongly depend on the respective application.

In the case of 2D imagery, features typically result from specific visual characteristics and are therefore referred to as *visual features*. Such visual features are of utmost importance in photogrammetry, remote sensing, computer vision and robotics, since they significantly facilitate a huge variety of crucial tasks reaching from image registration over image retrieval, object detection, object recognition, object tracking and navigation of autonomous vehicles to scene reconstruction or scene interpretation.

© Springer International Publishing Switzerland 2016
M. Weinmann, *Reconstruction and Analysis of 3D Scenes*,
DOI 10.1007/978-3-319-29246-5_3

In order to find a suitable definition for such a general expression, we may refer to the *Designs Act 2003* [11], where a *visual feature*, in relation to a product, includes the shape, configuration, pattern and ornamentation of the product and a *design*, in relation to a product, then means the overall appearance of the product resulting from one or more visual features of the product. In this context, it is explicitly pointed out that a visual feature may serve a functional purpose, but not necessarily, and that special attributes like the feel of the product and the materials used in the product are excluded. Transferring this rather general definition to image processing, a visual feature may be described as a special property of an image as a whole or an object within the image and it may either be a local property or a global characteristic of the image [41]. As a consequence, a visual feature may be defined in terms of intensity, shape, size, orientation, or texture.

In the case of 3D point cloud data, features typically result from specific geometric characteristics of the global or local 3D structure and therefore represent *geometric features*. More specifically, a geometric feature may be described as a special property of a 3D structure as a whole or only a part of the 3D structure. As a consequence, a geometric feature may be defined in terms of point attributes, shape, size, orientation, or roughness.

3.2   2D Feature Extraction

Generally, images acquired with a digital camera provide a discrete 2D projection of intensity measurements arranged on a regular image raster and therefore visual information is represented on a regular 2D grid of single picture elements referred to as *pixels*. Consequently, a variety of image processing techniques aims to process and analyze such images in order to describe and understand their content. For this purpose, numerous types of visual features have been introduced throughout the past decades, and many investigations have been carried out which clearly demonstrate that such features significantly alleviate common tasks in photogrammetry, remote sensing, computer vision, and robotics (e.g., image registration, image retrieval, object detection, object recognition, object tracking, navigation of autonomous vehicles, scene reconstruction, or scene interpretation). Surprisingly, approaches for extracting such features representing important visual details of an image even allow a prediction of image memorability [19]. In contrast to detailed considerations concerning feature extraction from images [24, 25, 36], we only summarize the main ideas based on our recent survey on visual features [41] which focuses on the categorization of different types of visual features, the link of all these feature types to a common source in visual perception, and the extension from single images to image sequences. Considering an image providing a variety of visual features (e.g., as the image provided in Fig. 3.1), we may easily reflect the main ideas of the different feature types covering overall appearance, pixel attributes, texture, shape and local features in the following subsections. Thereby, we take into account that, for being able to recognize the same

Fig. 3.1 An image revealing different types of visual features ("The Forest Has Eyes" © Bev Doolittle/The Greenwich Workshop, Inc.). Note that contextual information in the form of short- and mid-range dependencies causes illusions in the form of face-like structures and thus clearly indicates that an image might be more than the sum of its parts

or similar information in a different image, we do not only have to detect the features but also assign these features a feature description encapsulating the respective image properties.

3.2.1 Overall Appearance

Depending on the respective application, it may be desirable to describe an image as a whole by deriving a *holistic representation* of this image (e.g., for scene recognition). In this regard, fundamental properties of an image may be defined as the mean value or the standard deviation of intensity values, but also as the respective skewness, kurtosis, or entropy. Furthermore, the distribution of intensity values may be exploited which may for instance be described via histograms representing the occurrence of single intensity values while discarding their spatial arrangement. A consideration of the spatial arrangement of intensity values may involve moment theory [12], where early approaches are based on the idea of exploiting Moment Invariants [18] and modified variants which are for instance represented by Zernike Moment Invariants [33, 34] or Orthogonal Fourier–Mellin Moments [30]. For the latter two approaches, those moments with a lower order capture the low-frequency information and thus the coarse structure of an image, whereas the moments with a higher order capture the high-frequency information and thus image details. As a consequence, the image may be represented by a superposition of such moments and, depending on the order up to which moments are considered, a more or less accurate reconstruction of the image may be achieved with the involved moments. Instead of using moments, a further approach based on a coarse-to-fine image representation is represented by the discrete cosine transform [1] which has proven to be well-suited for image compression.

Among the more recent approaches, the GIST descriptor [26] focuses on obtaining the *gist*, an abstract representation of the scene which spontaneously activates memory representations of scene categories. More specifically, the GIST descriptor directly addresses perceptual properties represented by naturalness, openness, roughness, ruggedness, and expansion which are meaningful to human observers.

3.2.2   Pixel Attributes

Instead of describing an image as a whole, we may also focus on describing an observed image on the smallest possible scale, i.e., on pixel level, and exploit the fact that the pixels are assigned intensity measurements. Depending on the utilized device, the acquired intensity information may be represented in the form of gray-valued images where each pixel is assigned an intensity value, or in the form of color images where each pixel is assigned three values corresponding to the respective intensity values in the red (R), green (G), and blue (B) channels. For both types of images, intensity information represents a perceptual attribute which might be important to detect, segment or distinguish objects in an image or image regions with different semantic content. Thus, *intensity* represents a visual feature.

Due to technological advancements in the last decades, a variety of further sensor types relies on exploiting a discrete 2D projection of intensity measurements onto a regular image raster. Thereby, the intensity measurements may refer to different spectral bands and may hence even be beyond the visible spectrum. This is for instance exploited by thermal cameras providing information in the thermal infrared domain, and by satellite systems exploiting multispectral sensors for high-resolution satellite earth observation, e.g., IKONOS or QuickBird with four spectral bands covering the blue, green, red, and near-infrared spectra, or Landsat-7 with eight spectral bands covering the visible, near-infrared, shortwave infrared, and thermal infrared spectra [40].

However, other types of information may also be represented on a discrete image raster, e.g., information in the form of range images or depth images. Thus, intensity information may not only arise from measured spectral information, but also from a representation of data associated to each pixel as for instance shown in Fig. 3.2. For such a representation, an appropriate mapping has to be either defined or derived.

3.2.3   Texture

In contrast to a global consideration of the spatial arrangement of intensity values via moment theory, a local consideration of the spatial arrangement of intensity values

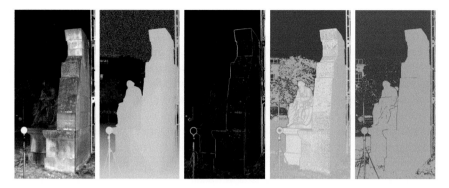

Fig. 3.2 Exemplary images with different types of information [42]: intensity image, range image, edge image, image encoding the local variation of range values, and binary image based on thresholding the local variation of range values (from *left* to *right*)

may facilitate a separation of local image regions corresponding to different objects or different semantic content. Thereby, we take into account that this spatial arrangement of intensity values may exhibit image regions of more or less homogeneity which, in turn, is often indicated by spatially repetitive (micro)structures across variations of scale and orientation. Consequently, the level of homogeneity which is commonly referred to as *texture* [14] may be considered as visual feature.

In general, texture may be defined as a function of the spatial variation of intensity values within an image [9]. This definition encapsulates contextual information involving intensity values within a local neighborhood, their spatial arrangement and different scales of resolution. In order to describe texture, Gabor filters [35] represented by sinusoidal plane waves within elliptical 2D Gaussian envelopes are often used, whereby a bank of Gabor filters at different scales and orientations is typically applied. Further commonly applied filter banks are represented by steerable filters [13], Leung and Malik filters [21], Schmid filters [29], and MR8 filters [38]. For more details on texture analysis and recent effort addressing material recognition, we refer to [8, 16, 37, 43, 44, 47, 53].

3.2.4 Shape

Early investigations on feature extraction already revealed that information is not uniformly distributed across the whole image, but rather concentrated along contours and particularly at points on a contour where its direction changes [4, 14]. In this context, a contour typically arises from an abrupt change in intensity or texture which, in turn, is likely to correspond to different materials, varying surface orientation or discontinuities with respect to illumination, distance, or reflectance behavior.

In order to detect respective contours and thus 2D shapes, standard approaches rely on convolving the image with a filter mask, where the latter is typically defined by the Roberts operator [28], the Prewitt operator [27], or the Sobel operator [32]. Since these approaches do not account for finding closed contours, a more sophisticated approach to contour detection has been presented with the Canny operator [7] which involves additional effort in order to recover closed contours.

In the case where the presence of specific contours may be expected in images, it is possible to locate respective shapes by involving prior knowledge about these expected contours, particularly if the considered contours may be assigned a contour parameterization by exploiting a mathematical model. For instance, in order to detect straight lines or line segments, the Hough transform [17] still represents one of the most commonly applied approaches. Since expected contours may not only consist of straight lines but also of further geometric primitives, the Hough transform has been adapted to circle and ellipse detection, and an extension to arbitrary shapes has been presented with the generalized Hough transform [5].

Once 2D shapes have been extracted, it is desirable to derive a suitable description. Respective approaches may generally be categorized into two groups [52]: Contour-based approaches either exploit global properties of a contour (e.g., perimeter, compactness, eccentricity, shape signature, Fourier descriptors, or wavelet descriptors) or structural properties (e.g., chain codes, polygonal approximations, splines, or boundary moments). In contrast, region-based approaches not only focus on contours but also account for the region within a closed contour. Consequently, they may either address global properties (e.g., size, orientation, area, eccentricity, or moments) or structural properties (e.g., convex hull, medial axis, or core).

3.2.5   Local Features

Stating that information is not uniformly distributed across the whole image, but rather concentrated along contours and particularly at points on a contour where its direction changes [4, 14], early investigations on feature extraction already concluded that highly informative image points are located at corners or blobs in an image [4] which, thus, address local characteristics of an image and are therefore commonly referred to as *local features*. In a more general definition, a local feature represents a pattern varying from its spatial neighborhood [36], where the variation may be due to a change of one or more image properties such as color, intensity or texture. Accordingly, we may not only define image corners and image blobs as local features, but also specific edge elements (*edgels*) or small distinctive image regions.

Typically, approaches for extracting local features may address *feature detection* in order to adequately retrieve the respective image locations and *feature description* in order to appropriately represent the local image characteristics. As a consequence, a *feature detector* only yields distinctive image locations without a feature representation which are therefore commonly referred to as *interest points* or *keypoints*, whereas the desirable representation for each distinctive image location is derived with a *feature descriptor* which, in turn, is commonly referred to as *interest point descriptor* or *keypoint descriptor*. Since the representation of a feature with a descriptor is relatively independent from the applied feature detector [10], feature detection and feature description may also be treated separately, and a large variety of approaches addressing both tasks has been presented in literature.

In comparison to other types of visual features, local features reveal interesting properties. Since such local features (i) may be extracted in a highly efficient manner, (ii) are accurately localized, (iii) remain stable over reasonably varying viewpoints[1] and (iv) allow an individual identification, they are well-suited for a large variety of applications such as object detection, object recognition, autonomous navigation and exploration, image and video retrieval, image registration or the reconstruction, interpretation and understanding of scenes [22, 36, 41].

3.3 3D Feature Extraction

In analogy to feature extraction from 2D imagery, we may define respective feature types for 3D point cloud data. For this purpose, we may directly transfer the basic ideas of different feature types from 2D space to 3D space. In this regard, however, we have to take into account the difference in the neighborhood definition. Whereas image neighborhoods rely on the regular raster and may therefore easily and efficiently be derived, we will typically not face 3D point clouds where the single 3D points represent a regular 3D raster. Considering a typical representation in the form of a 3D point cloud (e.g., the 3D point cloud shown in Fig. 3.3), we may easily reflect the main ideas of the different feature types covering point attributes, shape and local features in the following subsections.

3.3.1 Point Attributes

Depending on the acquisition or generation of 3D point cloud data, particular attributes such as *intensity information* (either in the form of gray values or in the form of color) may be available which, in turn, may be defined as features. For instance, modern laser scanners (e.g., Leica HDS6000 or Riegl LMS-Z360i) and range cameras (e.g., PMD[vision] CamCube 2.0 or MESA Imaging SR4000) provide the capability to acquire intensity information. The Microsoft Kinect even allows to acquire intensity information in the red (R), green (G), and blue (B) channels and thus color information for each 3D point. Similarly, 3D point clouds generated via Multi-View Stereo (MVS) methods may carry intensity information in the form of gray values

---

[1]Considering the function of rotation in depth of a plane away from a viewer, some of these features are reported to be stable up to a change of 50° in viewpoint [23].

Fig. 3.3 Exemplary 3D point cloud revealing different types of features: the 3D point cloud consists of > 20 M 3D points representing an urban environment

if gray-valued images are used for 3D reconstruction or in the form of color values if color images are used for 3D reconstruction.

In the case where one device might not be sufficient to capture the desired information about a scene, different devices may be involved and this results in the need for a co-registration of the captured data. For instance, a digital camera may be mounted on a laser scanner [2] in order to enrich the 3D point cloud data with color information. Furthermore, an enrichment of 3D point cloud data may also focus on a respective mapping of thermal information onto the 3D point cloud data [45, 46] which we will address in Chap. 5.

Finally, we may take into account that it is also possible to define arbitrary attributes for each 3D point. Accordingly, we may associate specific attributes such as the spatial 3D coordinates, the components of the local normal vector, the scale or possibly even abstract information to each 3D point for which some examples are shown in Fig. 3.4.

3.3.2   Shape

When describing 3D shapes, we may focus on the description of either single 3D objects or whole 3D scenes and we may face very different requirements in both cases. While single 3D objects or 3D object models typically provide *closed 3D shapes*, a respective geometric description may easily be derived from global geometric properties such as size, orientation, volume, or eccentricity. Furthermore, we may derive a representation of the volumetric distribution via 3D moments, boundary moments (i.e., surface moments) or shape signatures. However, such closed 3D

Fig. 3.4 Exemplary 3D point clouds exhibiting different types of features [42]: the depicted point clouds encode representations with respect to color, gray values, height information, and a measure of planarity (from *left* to *right*)

shapes are typically not given in 3D point clouds representing whole 3D scenes with urban structures and vegetation, even if such an environment is decomposed into smaller segments. For this reason, the majority of approaches exploit information about the local surface structure of objects in order to adequately represent *arbitrary 3D shapes*. In this regard, the local surface structure of a 3D point is described by considering the spatial arrangement of 3D points within a local neighborhood, and all 3D points within the neighborhood are exploited in order to derive a suitable description.

3.3.3 Local Features

Transferring the idea of local features from 2D imagery to 3D point cloud data, we may consider 3D locations varying from their spatial 3D neighborhood as *local features*. Note that, in this case, the variation may be due to a change with respect to the local 3D structure, and local features therefore typically correspond to corners or blobs in 3D space. In order to extract such local features from 3D point cloud data, the general approach only exploiting spatial 3D geometry is addressed by *3D feature extraction*. When using range cameras, however, the spatial 3D information may also be represented in the form of a range image, where range information is available for discrete points on a regular 2D grid. Consequently, feature extraction may also rely on the range image and thus be referred to as *2.5D feature extraction*.

While the extraction of local 3D features and local 2.5D features mainly differs in the involved neighborhood definitions, respective approaches may generally address *feature detection* which aims at finding distinctive 3D locations (which are referred to as *3D interest points* or *3D keypoints*) and *feature description* which focuses on deriving compact and appropriate representations in order to adequately describe the characteristics of the local 3D structure. In the last decade, a variety of 3D feature detectors and 3D feature descriptors has been presented in literature. For more details, we refer to recent surveys [42, 51].

3.4   Discussion

Following our findings on 2D feature extraction, we are able to detect and describe a variety of visual characteristics when considering 2D data represented in the form of images. However, the different feature types covering overall appearance, pixel attributes, texture, shape, and local features are only a more or less compact representation of specific properties contained in the data and thus only suitable to encapsulate a very limited amount of information about an observed scene. Furthermore, we should not forget that the suitability of a feature may generally strongly depend on the respective application. For instance, it may be very hard or even impossible for a fully automated system to understand the image depicted in Fig. 3.1 only by exploiting such visual features, while it may be much easier to detect the person in the scene as well as the two horses. In this regard, we may also observe that it is not only the objects themselves but also their spatial arrangement which characterize a scene. For us humans, specific arrangements of objects may even result in illusions arising from *contextual information* in the form of short- and mid-range dependencies. While we are able to detect many faces in this given image since human perception is biased and trained to see faces, our mind tells us that it might be rather by occasion and not too relevant for describing the happening in the scene. Note that such contextual information may thus be considered as one of the reasons why we may clearly agree with the statement of Aristotle that "the whole is more than the sum of its parts" [3].

Regarding feature extraction, it should also be taken into account that the different types of visual features have been derived from numerous investigations on human visual perception, where different psychological theories have been presented and early studies may even be traced back to ancient Greek theories on how vision is carried out. We have focused on this issue in our recent survey on the extraction of visual features [41] and will now only briefly explain those ideas with the most significant impact on recent research.

Among the most prominent psychological theories, the theory of *Structuralism* represents a very simple and rather intuitive principle claiming that perception is based on a large number of sensory atoms measuring color at specific locations within the field-of-view. While this principle could for instance motivate the use of pixel attributes or—when considering small image neighborhoods—the use of local structural properties of an image, it becomes apparent that visual perception not only involves local image characteristics such as lines, angles, colors, or shape, but also grouping processes for what has been detected with the senses and the respective interpretation which, in turn, would be in accordance with the aforementioned statement of Aristotle. Consequently, the theory of *Holism* has been presented which claims that the whole cannot be represented by utilizing the sum of its parts alone due to relations between the parts, but rather "as a unity of parts which is so close and intense as to be more than the sum of its parts; which not only gives a particular conformation or structure to the parts but so relates and determines them in their synthesis that their functions are altered; the synthesis affects and determines the parts, so that they function towards the *whole*; and the whole and the parts therefore reciprocally influence and determine each other, and appear more or less to merge their individual characters: the whole is in the parts and the parts are in the whole, and this synthesis of whole and parts is reflected in the holistic character of the functions of the parts as well as of the whole" [31]. A further psychological theory has been presented with *Gestalt Theory* [39, 48], where the *gestalt* may be seen as a German synonym for shape, form, figure, or configuration. The main ideas of Gestalt Theory may be summarized as stated in 1924 by explaining that "there are entities where the behavior of the whole cannot be derived from its individual elements nor from the way these elements fit together; rather the opposite is true: the properties of any of the parts are determined by the intrinsic structural laws of the whole" [50]. These "intrinsic structural laws of the whole" correspond to relations between the parts and, according to seminal work [20, 49], the visual system exploits different principles for automatically grouping elements of an image into patterns such as proximity, similarity, closure, symmetry, common fate (i.e., common motion), and continuity. In [14], further principles have been proposed in the form of homogeneity, which has been introduced as texture, and in the form of contour which represents a physical edge caused by an abrupt change in color or texture.

However, describing an image as a whole might be very complex and therefore, the theory of *Reductionism* became more and more important which goes back to ideas of e.g., René Descartes according to which the understanding of a system as a whole may be achieved by attempting to explain its constituent parts (i.e., the theory of *Atomism* according to which the examined parts can be put together again in order to make the whole) and their interactions (i.e., some kind of context). Accordingly, one could claim that an image can be explained by reduction to its fundamental parts. Yet, following René Descartes, visual perception is based on representations of reality which have been formed within ourselves [6]. In contrast to such an indirect perception, the *Theory of Ecological Optics* [14, 15] has been presented which focuses on an active and direct perception of visual information, and where—by involving basic principles of Gestalt Theory as well as aspects of information theory—a modified set of 10 principles addressing varieties of continuous regularity, discontinuous regularity or recurrence, proximity, and situations involving interaction has for instance been proposed in [4]. These principles, in turn, arise from three main observations. The first one focuses on the finding that information is not uniformly distributed but rather concentrated along contours and particularly at points on a contour at which its direction changes, i.e., corners. The second observation addresses the fact that contours are caused by changes of homogeneity with respect to color or texture, and the third observation indicates that the degree of homogeneity, i.e., texture, can be considered as a characteristic as well. These findings, in turn, lead to several types of visual features in imagery such as corners (i.e., local features), contours (i.e., shape), color/intensity (i.e., pixel attributes) or homogeneity (i.e., texture) which have been considered in different subsections.

Our considerations reveal that the different psychological theories cover important ideas, and—with a more detailed literature review—one may agree that the cognitive and computational approaches to visual perception are mainly based on combining descriptive principles already proposed in Gestalt Theory and the Theory of Ecological Optics with a model of perceptual processing.

While the complexity of images of an observed scene may be quite high due to the scene content, a high complexity of 3D point cloud data may already be caused by an irregular sampling and a varying point density as shown in Fig. 3.3. In order to describe such a scene, we may extract different feature types covering point attributes, shape and local features which, in analogy to visual features, only provide a more or less compact representation of specific scene properties and thus only encapsulate a very limited amount of information about the scene. A fully automated interpretation of this 3D point cloud however remains challenging. Even for us humans, it is not so easy to adequately describe this 3D point cloud. Some of the given shapes indicate building façades. Together with the center of the 3D point cloud which possibly represents streets and/or pavements, these building façades form an urban environment. In contrast to geometric properties, however, the color encoding may hardly be interpretable for people who are unfamiliar with 3D point cloud processing. Even a more experienced user may need some time to identify that the 3D point clouds of different color represent different parts of the scene. The high overlap and the circles visible in the center of the scene indicate the use of a terrestrial laser scanner for data acquisition and, finally, that the whole 3D point cloud may represent the result of point cloud registration, i.e., the alignment of several 3D point clouds in a common coordinate frame.

3.5   Conclusions

In this chapter, we have briefly reflected fundamentals of feature extraction and thereby focused on feature extraction from 2D imagery and feature extraction from 3D point cloud data. These fundamentals have become the basis and core of a plethora of applications in photogrammetry, remote sensing, computer vision and robotics, and they also provide the basis for the subsequent chapters focusing on point cloud registration (Chap. 4), co-registration of 3D and 2D data (Chap. 5) and 3D scene analysis (Chap. 6). Instead of going too much into detail, however, we have only reflected the main ideas which allow us to conclude that the suitability of different approaches strongly depends on the respective task or application, and we will therefore have to motivate and consider those methods involved in our work in the respective chapters. For comprehensive and more detailed considerations on feature extraction—which are beyond the scope of this book—we refer to our recent surveys on 2D and 3D feature extraction [41, 42].

References

1. Ahmed N, Natarajan T, Rao KR (1974) Discrete cosine transform. IEEE Trans Comput C–23(1):90–93
2. Al-Manasir K, Fraser CS (2006) Registration of terrestrial laser scanner data using imagery. Photogramm Rec 21(115):255–268
3. Aristotle (384–322 B.C.) Metaphysica
4. Attneave F (1954) Some informational aspects of visual perception. Psychol Rev 61(3):183–193
5. Ballard DH (1981) Generalizing the Hough transform to detect arbitrary shapes. Pattern Recognit 13(2):111–122
6. Braund MJ (2008) The structures of perception: an ecological perspective. Kritike 2(1):123–144
7. Canny J (1986) A computational approach to edge detection. IEEE Trans Pattern Anal Mach Intell 8(6):679–698
8. Caputo B, Hayman E, Fritz M, Eklundh J-O (2010) Classifying materials in the real world. Image Vis Comput 28(1):150–163
9. Chen CH, Pau LF, Wang PSP (1998) Handbook of pattern recognition and computer vision, 2nd edn. World Scientific Publishing, Singapore
10. Dahl AL, Aanæs H, Pedersen KS (2011) Finding the best feature detector-descriptor combination. In: Proceedings of the IEEE international conference on robotics and automation, pp 318–325
11. Designs Act (2003) Office of legislative drafting and publishing. Attorney-General's Department, Canberra, Australia
12. Flusser J, Suk T, Zitova B (2009) Moments and moment invariants in pattern recognition. Wiley, Chichester
13. Freeman W, Adelson EH (1991) The design and use of steerable filters. IEEE Trans Pattern Anal Mach Intell 13(9):891–906
14. Gibson JJ (1950) The perception of the visual world. Houghton Mifflin, Boston
15. Gibson JJ (1961) Ecological optics. Vis Res 1(3–4):253–262
16. Haralick RM (1979) Statistical and structural approaches to texture. Proc IEEE 67(5):786–804
17. Hough PVC (1962) Method and means for recognizing complex patterns. US Patent 3069654
18. Hu M-K (1962) Visual pattern recognition by moment invariants. IRE Trans Inf Theory 8(2):179–187
19. Isola P, Xiao J, Torralba A, Oliva A (2011) What makes an image memorable? In: Proceedings of the IEEE computer society conference on computer vision and pattern recognition, pp 145–152
20. Koffka K (1935) Principles of gestalt psychology. Harcourt, Brace & World, New York
21. Leung T, Malik J (1999) Recognizing surfaces using three-dimensional textons. In: Proceedings of the IEEE international conference on computer vision, vol 2, pp 1010–1017
22. Li J, Allinson NM (2008) A comprehensive review of current local features for computer vision. Neurocomputing 71(10–12):1771–1787
23. Lowe DG (2004) Distinctive image features from scale-invariant keypoints. Int J Comput Vis 60(2):91–110
24. Ma Y, Soatto S, Košecká J, Sastry SS (2005) An invitation to 3-D vision: from images to geometric models. Springer, New York
25. Nixon MS, Aguado AS (2008) Feature extraction and image processing, 2nd edn. Academic Press, Oxford
26. Oliva A, Torralba A (2001) Modeling the shape of the scene: a holistic representation of the spatial envelope. Int J Comput Vis 42(3):145–175
27. Prewitt JMS, Mendelsohn ML (1966) The analysis of cell images. Ann NY Acad Sci 128(3):1035–1053

28. Roberts LG (1965) Machine perception of three-dimensional solids. In: Tippett J, Berkowitz D, Clapp L, Koester C, Vanderburgh A (eds) Optical and electro-optical information processing. MIT Press, Cambridge, pp 159–197
29. Schmid C (2001) Constructing models for content-based image retrieval. In: Proceedings of the IEEE computer society conference on computer vision and pattern recognition, vol 2, pp 39–45
30. Sheng Y, Shen L (1994) Orthogonal Fourier–Mellin moments for invariant pattern recognition. J Opt Soc Am A 11(6):1748–1757
31. Smuts J (1926) Holism and evolution. The MacMillan Company, New York
32. Sobel IE (1970) Camera models and machine perception. PhD thesis, Stanford University, Stanford, USA
33. Teague MR (1980) Image analysis via the general theory of moments. J Opt Soc Am 70(8):920–930
34. Teh C-H, Chin RT (1988) On image analysis by the methods of moments. IEEE Trans Pattern Anal Mach Intell 10(4):496–513
35. Turner MR (1986) Texture discrimination by Gabor functions. Biol Cybern 55(2):71–82
36. Tuytelaars T, Mikolajczyk K (2008) Local invariant feature detectors: a survey. Found Trends Comput Graph Vis 3(3):177–280
37. Van Gool L, Dewaele P, Oosterlinck A (1985) Texture analysis Anno 1983. Comput Vis Graph Image Process 29(3):336–357
38. Varma M, Zisserman A (2002) Classifying images of materials: achieving viewpoint and illumination independence. In: Proceedings of the European conference on computer vision, vol 3, pp 255–271
39. von Ehrenfels C (1890) Über Gestaltqualitäten. Vierteljahrsschrift für wissenschaftliche Philosophie 14:249–292
40. Weidner U (2005) Remote sensing systems—An overview focussing on environmental applications. In: Proceedings of the EnviroInfo 2005 workshop on tools for emergencies and disaster management, pp 1–8
41. Weinmann M (2013) Visual features—From early concepts to modern computer vision. In: Farinella GM, Battiato S, Cipolla R (eds) Advanced topics in computer vision. Advances in computer vision and pattern recognition. Springer, London, pp 1–34
42. Weinmann M (2016) Feature extraction from images and point clouds: fundamentals, advances and recent trends. Whittles Publishing, Dunbeath (to appear)
43. Weinmann Mi, Klein R (2015) A short survey on optical material recognition. In: Proceedings of the eurographics workshop on material appearance modeling: issues and acquisition, pp 35–42
44. Weinmann Mi, Klein R (2015) Advances in geometry and reflectance acquisition (course notes). In: Proceedings of the SIGGRAPH Asia 2015 courses, pp 1:1–1:71
45. Weinmann M, Hoegner L, Leitloff J, Stilla U, Hinz S, Jutzi B (2012) Fusing passive and active sensed images to gain infrared-textured 3D models. Int Arch Photogramm Remote Sens Spat Inf Sci XXXIX-B1:71–76
46. Weinmann M, Leitloff J, Hoegner L, Jutzi B, Stilla U, Hinz S (2014) Thermal 3D mapping for object detection in dynamic scenes. ISPRS Ann Photogramm Remote Sens Spat Inf Sci II-1:53–60
47. Weinmann Mi, Gall J, Klein R (2014) Material classification based on training data synthesized using a BTF database. In: Proceedings of the European conference on computer vision, vol III, pp 156–171
48. Wertheimer M (1912) Experimentelle Studien über das Sehen von Bewegung. Zeitschrift für Psychologie 61(1):161–265
49. Wertheimer M (1923) Untersuchungen zur Lehre von der Gestalt. Psychologische Forschung 4:301–350
50. Westheimer G (1999) Gestalt theory reconfigured: Max Wertheimer's anticipation of recent developments in visual neuroscience. Perception 28(1):5–15

51. Yu T-H, Woodford OJ, Cipolla R (2013) A performance evaluation of volumetric 3D interest point detectors. Int J Comput Vis 102(1–3):180–197
52. Zhang D, Lu G (2004) Review of shape representation and description techniques. Pattern Recognit 37(1):1–19
53. Zhang J, Tan T (2002) Brief review of invariant texture analysis methods. Pattern Recognit 35(3):735–747

References

Chapter 4
Point Cloud Registration

When focusing on an automated digitization, reconstruction, interpretation, and understanding of an observed scene, a crucial task consists of obtaining an adequate representation of the considered scene. Particularly a geometric representation, e.g., in the form of a 3D point cloud as densely sampled and accurately described counterpart of physical object surfaces or in the form of a mesh as collection of vertices, edges, and faces as approximation of the shape of a polyhedral object, facilitates a variety of applications such as object recognition, navigation of autonomous vehicles, guidance of people in wheelchairs, or a semantic scene interpretation. Since the vertices of a mesh also represent a 3D point cloud, we focus on the consideration of 3D point clouds in the scope of our work. Yet for both cases, a dense and accurate acquisition of the considered scene is desirable which, in turn, implies that data acquisition from a single location might not be sufficient in order to obtain a dense and (almost) complete 3D acquisition of interesting parts of the considered scene. As a consequence, respective data has to be collected from different locations. While modern scanning devices allow a direct acquisition of single scans representing dense and accurate 3D point cloud data, a data acquisition from different locations results in the fact that each 3D point cloud is only measured with respect to the respective location and orientation of the scanning device. Consequently, approaches for an appropriate alignment of several 3D point clouds in a common coordinate frame—which is referred to as *point cloud registration*—are required.

In this chapter, we focus on keypoint-based point cloud registration which has proven to be among the most efficient strategies for aligning pairs of overlapping scans. We present a novel and fully automated framework which consists of six components addressing (i) the generation of 2D image representations in the form of range and intensity images, (ii) point quality assessment, (iii) feature extraction and matching, (iv) the forward projection of 2D keypoints to 3D space, (v) correspondence weighting, and (vi) point cloud registration. For the respective components, we take into account different approaches and our main contributions address the issue of how to increase the robustness, efficiency, and accuracy of point cloud registration by either introducing further constraints (e.g., addressing a correspondence weighting based on point quality measures) or replacing commonly applied approaches by

© Springer International Publishing Switzerland 2016
M. Weinmann, *Reconstruction and Analysis of 3D Scenes*,
DOI 10.1007/978-3-319-29246-5_4

more promising alternatives. The latter may not only address the involved strategy for point cloud registration, but also the involved approaches for feature extraction and matching. In a detailed evaluation, we demonstrate that, instead of directly aligning sets of corresponding 3D points, a transfer of the task of point cloud registration to the task of solving the Perspective-n-Point (PnP) problem or to the task of finding the relative orientation between sets of bearing vectors offers great potential for future research. Furthermore, our results clearly reveal that the further consideration of both a correspondence weighting based on point quality measures and a selection of an appropriate feature detector–descriptor combination may result in significant advantages with respect to robustness, efficiency, and accuracy.

Our framework has successfully undergone peer review and our main contributions address a projective scan matching technique [100, 102–105], the impact of a binary weighting of feature correspondences by simply removing unreliable ones based on different point quality measures [100, 103] and the influence of a numerical weighting of feature correspondences based on point quality assessment [101, 102, 106]. Further contributions are represented by an omnidirectional scan matching technique [95] and an evaluation of the respective impact of different feature detector–descriptor combinations on the results of point cloud registration [95].

After providing a brief motivation for point cloud registration (Sect. 4.1), we reflect related work and thereby focus on different types of features which may facilitate the task of point cloud registration (Sect. 4.2). In this context, particularly local features offer many advantages and we therefore briefly explain the main ideas of the most commonly applied approaches for aligning 3D point clouds based on the use of keypoints. Subsequently, we present our framework and explain the different components in detail (Sect. 4.3). In order to demonstrate the performance of our framework, we provide an extensive experimental evaluation on a publicly available benchmark dataset (Sect. 4.4) and discuss the derived results (Sect. 4.5). Finally, we provide concluding remarks and suggestions for future work (Sect. 4.6).

4.1 Motivation and Contributions

The faithful 3D reconstruction of indoor and outdoor environments represents a topic of great interest in photogrammetry, remote sensing, computer vision, and robotics. While a variety of devices allows to acquire an appropriate representation of local object surfaces in the form of 3D point cloud data as measured counterpart of physical surfaces, terrestrial laser scanning (TLS) systems provide dense and accurate 3D point cloud data for the local environment and they may also reliably measure distances of several tens of meters. Due to these capabilities, such TLS systems are commonly used for applications such as city modeling, construction surveying, scene interpretation, urban accessibility analysis, or the digitization of cultural heritage objects. When using a TLS system, each captured TLS scan is represented in the form of a 3D point cloud consisting of a large number of scanned 3D points and, optionally, additional attributes for each 3D point such as color or intensity information. However, a TLS system represents a line-of-sight instrument and hence occlusions resulting from objects in the scene may be expected as well as a significant variation in point density between close and distant object surfaces (see e.g., Fig. 4.1). Thus, a single scan might not be sufficient in order to obtain a dense and (almost) complete 3D acquisition of interesting parts of a scene and, consequently, multiple scans have to be acquired from different locations. Since the spatial 3D information of each of the respective 3D point clouds is only measured with respect to the local coordinate frame of the scanning device, it is desirable to transfer all captured 3D point cloud data into a common coordinate frame and thus accurately align the single 3D point clouds. For this purpose, the respective transformations between the captured 3D point clouds have to be estimated and this process is commonly referred to as *point cloud registration, point set registration,* or *3D scan matching.*

The process of point cloud registration is typically divided into a *coarse registration* and a *fine registration*. Whereas the coarse registration only provides a coarse, but fast initial estimate of the transformation between two 3D point clouds, the fine registration aims at refining this initial estimate in order to recover the transformation as accurately as possible. For both steps, the proposed approaches may be categorized according to (i) the given data, (ii) the type of corresponding information, (iii) the applied strategy, and (iv) the methodology itself. Considering the captured data, it is meanwhile possible to exploit only spatial 3D geometry, to exploit spatial 3D geometry in combination with respective range images, or to exploit spatial 3D geometry in combination with respective intensity images. Corresponding information is typically represented in the form of corresponding points, lines, or surfaces. The strategy may either be an iterative or a noniterative scheme, and it may address the estimation

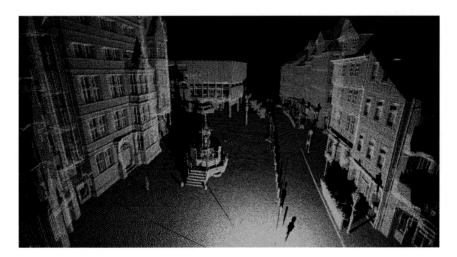

Fig. 4.1 Visualization of a 3D point cloud acquired with a terrestrial laser scanner: occlusions resulting from objects in the scene as well as a varying point density become clearly visible

of a rigid transformation preserving the distance between 3D points within the considered 3D point clouds or a nonrigid transformation allowing deformations of the considered 3D point clouds. Depending on the given data as well as the derived correspondences and the chosen strategy, the methodology may exploit a large variety of techniques. These techniques may also depend on the respective application and thus provide a trade-off between processing time, memory consumption, and registration accuracy. After fine registration, a subsequent *global registration* also known as *multiview registration* may optionally be carried out in order to globally optimize the alignment of all 3D point clouds by simultaneously minimizing registration errors with respect to all available scans. Valuable surveys address general criteria for point cloud registration [69], the most common techniques for coarse and fine registration [75], and the issue of rigid or nonrigid registration [84].

In this chapter, we address point cloud registration for data acquired with modern scanning devices such as terrestrial laser scanners and range cameras which allow to acquire spatial 3D information and to derive respective 2D image representations in the form of range and intensity images (see Sect. 2.4). For this purpose, we take into account that nowadays most approaches rely on exploiting specific features extracted from the acquired data instead of considering the complete 3D point clouds. Particularly, local features in the form of keypoints with respectively assigned keypoint descriptors have become a ubiquitous means for point cloud registration, since they may efficiently be extracted from either the acquired 3D point clouds or the respectively derived 2D image representations and furthermore allow a relatively robust feature matching which, in turn, significantly facilitates point cloud registration. Respective approaches which may be referred to as *keypoint-based point cloud registration approaches* have proven to be among the most efficient approaches for aligning pairs of overlapping scans. The main idea of such approaches consists of exploiting corresponding keypoints in order to derive sparse sets of corresponding 3D points. However, instead of directly exploiting these corresponding 3D points for point cloud registration by considering 3D cues in terms of how good corresponding 3D points fit together with respect to the Euclidean distance, we are interested in also considering 2D cues in terms of how good the 2D locations in the respective image representations fit together. As a result of taking into account both 3D and 2D cues which may principally rely on different strategies, we expect to achieve an improvement in accuracy for the registration results. Consequently, we investigate two promising strategies which relate the task of point cloud registration to (i) the task of solving the Perspective-n-Point (PnP) problem and (ii) the task of finding the relative orientation between sets of bearing vectors.

In summary, our main contribution consists of a fully automated, efficient, and robust framework which involves

- the consideration of different feature detector–descriptor combinations in order to extract local features from the intensity images derived for the respective scans,
- a novel approach for organizing TLS point clouds with respect to their similarity by deriving an appropriate scan order for a successive pair-wise registration scheme,

- the definition of a consistency check based on novel measures for assessing the quality of scanned 3D points in order to remove unreliable feature correspondences,
- the transfer of the task of point cloud registration to the task of solving the Perspective-n-Point (PnP) problem, and
- the transfer of the task of point cloud registration from the object space (i.e., the direct alignment of 3D point sets) to the observation space (i.e., the direct alignment of sets of bearing vectors).

4.2 Related Work

Nowadays, the most accurate alignment of TLS point clouds is still obtained via manipulating the observed scene by placing artificial markers which represent clearly demarcated corresponding points in different scans. Thus, such markers may easily be extracted either manually or automatically [3, 31]. Even though a good quality of the registration process is ensured, this procedure may however be rather time-consuming—particularly for a large number of scans—and it hence often tends to be intractable. Consequently, a fully automated registration of scans without using artificial markers is desirable. Such a fully automated registration may principally rely on very different types of features which we address in the following (Sect. 4.2.1). A deeper analysis of these features, in turn, reveals that specific features are particularly suited for an automatic registration of scans acquired with TLS systems and have therefore been involved for a variety of approaches (Sect. 4.2.2).

4.2.1 Feature Extraction

Generally, we may categorize approaches for point cloud registration with respect to the exploited data as done in the following subsections. Thereby, respective approaches may rely on geometric features (Sect. 4.2.1.1), on features extracted from range images (Sect. 4.2.1.2), or on features extracted from intensity images (Sect. 4.2.1.3). Since particularly keypoint detectors and descriptors may significantly facilitate point cloud registration, we will also briefly explain the main ideas in this regard (Sect. 4.2.1.4).

4.2.1.1 Geometric 3D Features

A variety of standard approaches only exploits spatial 3D geometry for point cloud registration. Among these techniques, the *Iterative Closest Point (ICP) algorithm* [18, 25] and its variants [4, 24, 32, 38, 59, 72] have become very popular and are meanwhile commonly used. In general, the ICP algorithm is based on iteratively minimizing the difference between two considered 3D point clouds, where the mean square error between 3D points of a 3D point cloud and the closest 3D points of another 3D point cloud is considered. However, the iterative scheme may be very time-consuming for large 3D point clouds. With *Least Squares 3D Surface Matching* (*LS3D*) [33], a further technique has been proposed which relies on minimizing the Euclidean distance between matched surfaces in the least squares sense. Both the ICP algorithm and the LS3D method focus on a local optimization and require a good initial alignment in order to converge to the correct transformation. For this reason, such techniques are commonly used for fine registration where a sufficiently good initial alignment is already available.

In order to avoid the need for a good initial alignment and furthermore increase computational efficiency, it is advisable to extract relevant information in the form of specific corresponding features from the considered 3D point clouds. Such relevant information may for instance be derived from the distribution of the 3D points within each 3D point cloud using the normal distributions transform (NDT) [19] either on 2D scan slices [21] or in 3D space [55]. If the presence of specific regular surfaces may be assumed as for instance within urban environments or scenes containing industrial installations, the detection of corresponding features may rely on specific geometric primitives such as planes [21, 38, 63, 85, 96] and/or even more complex primitives such as spheres, cylinders, and tori [67]. Particularly planar features are well-suited as they still allow a relatively robust registration in case of larger distances between the respective scan positions [21], and they may also be combined with the linear features arising from intersections of neighboring planar structures or the boundaries of planar structures [79]. A more sophisticated approach relying on the use of planes has for instance been presented in order to adequately align ALS point clouds by fitting planes to the data and removing those 3D points on objects which are susceptible to occlusions, shadows, etc. as well as those 3D points on objects with irregular shape while keeping only 3D points that are promising for a robust point cloud registration [38–40]. However, all these feature types representing specific geometric primitives encounter significant challenges in case of scene symmetry and they are not suited in scenes without regular surfaces.

Facing general scenes where we may not assume the presence of specific 3D shape primitives, features representing distinctive properties of local surface patches may be more appropriate and the least assumptions are made when focusing on point-like features. Such features may for instance be based on geometric curvature or normal vectors of the local surface [12, 43], on Point Feature Histograms [73, 74] or on the application of keypoint detectors in 3D space, e.g., a 3D Harris corner detector or a 3D Difference-of-Gaussians (DoG) detector [86–88]. However, in order to increase computational efficiency and furthermore account for the fact that—due to the use of a line-of-sight instrument with a specific angular resolution—a significant variation in point density may be expected, most of the proposed 3D interest point detectors are based on a voxelization of the scene and thus strongly depend on the selected voxel size.

4.2.1.2   Features Extracted from Range Images

Taking into account that the range information is acquired for points on a regular scan grid, we may easily derive an image representation in the form of range images and efficiently extract interest points from these range images, e.g., by applying a min-max algorithm [14], the Harris corner detector [80], the Laplacian-of-Gaussian (LoG) detector [81], or the Normal Aligned Radial Feature (NARF) detector [82]. While the design of an interest point detector may principally also account for finding keypoints on geometrically smooth 3D surfaces, such approaches generally require characteristic 3D structures in the scene which may not always be that well-distributed in larger distances to the scanning device. Furthermore, an approach has been presented which relies on features arising from a distinctive skyline [62], since the skyline shows extrema, flat regions or rising and falling edges when using a cylindrical projection model in order to sample the range information to panoramic range images covering 360° in the horizontal direction. The relative arrangement of such features is well-suited for coarsely aligning two scans, but the respective features strongly depend on the scene content, i.e., severe problems might arise if the skyline is less distinctive. Further 3D information, which may be projected onto a regular grid and thus exploited for point cloud registration, can be derived from the normal vectors of the respective surface. From these, Extended Gaussian Images (EGIs) [41] may be derived as well as surface orientation histograms approximating the EGI representation [27, 56].

4.2.1.3   Features Extracted from Intensity Images

Since modern scanning devices also allow to acquire intensity information on the discrete scan grid and thus intensity images which typically provide complementary information with a higher level of distinctiveness than range images [77], most of the proposed approaches for point cloud registration nowadays exploit spatial 3D geometry in combination with intensity images. In particular, such intensity images allow to detect reliable correspondences between visual features, whereas range images typically suffer from the occurrence of periodic shapes in man-made environments which exhibit similar and thus often ambiguous visual features. As already explained in Chap. 3, local features are well-suited for reliably detecting correspondences between distinctive features of intensity images due to their accurate localization, their stability over reasonably varying viewpoints, the possibility of individual identification, and the efficient algorithms for extracting them [93, 99]. Such local features are characterized by a *keypoint* representing a respective image location and a *keypoint descriptor* encapsulating a representation of the local image characteristics. Accordingly, many approaches for point cloud registration involve features in the form of keypoints extracted from intensity images derived from reflectance data [6, 20, 45, 97] or co-registered camera images [5, 13, 17, 107]. The respective forward projection of such 2D keypoints according to the corresponding range information allows to derive sparse sets of (almost) identical 3D points and thus significantly alleviates point cloud registration while improving computational efficiency.

4.2.1.4 Keypoint Extraction and Matching

Generally, different types of visual features may be extracted from images in order to detect corresponding image contents [99]. However, local features such as image corners, image blobs, or small image regions offer significant advantages. Since such local features (i) may be extracted very efficiently, (ii) are accurately localized, (iii) remain stable over reasonably varying viewpoints, and (iv) allow an individual identification, they are well-suited for a large variety of applications such as object detection, object recognition, object tracking, autonomous navigation and exploration, image and video retrieval, image registration or the reconstruction, interpretation, and understanding of scenes [93, 99]. As outlined in Sect. 3.2.5, the extraction of local features consists of two steps which are represented by *feature detection* and *feature description*.

For feature detection, corner detectors such as the Harris corner detector [36] or the Features from Accelerated Segment Test (FAST) detector [70] are widely used. The detection of blob-like structures is typically solved with a Difference-of-Gaussian (DoG) detector which is integrated in the Scale Invariant Feature Transform (SIFT) [53, 54], or a Determinant-of-Hessian (DoH) detector which is the basis for deriving Speeded-Up Robust Features (SURF) [15, 16]. Distinctive image regions are for instance detected with a Maximally Stable Extremal Region (MSER) detector [57]. Accounting for non-incremental changes between images with similar content and thus possibly significant changes in scale, the use of a scale-space representation as introduced for the SIFT and SURF detectors is inevitable. While the SIFT and SURF detectors rely on a Gaussian scale-space, the use of a nonlinear scale-space has been proposed for detecting KAZE features [7] or Accelerated KAZE (A-KAZE) features [8].

For feature description, the main idea consists of deriving keypoint descriptors that allow to discriminate the extracted keypoints very well. Being inspired by investigations on biological vision, the descriptor presented as second part of the Scale Invariant Feature Transform (SIFT) [53, 54] is one of the first and still one of the most powerful feature descriptors. Since, for applications focusing on computational efficiency, the main limitation of deriving SIFT descriptors consists of the computational effort, more efficient descriptors have been presented with the Speeded-Up Robust Features (SURF) descriptor [15, 16] and the DAISY descriptor [89, 90]. In contrast to these descriptors consisting of a vector representation encapsulating floating numbers, a significant speedup is typically achieved by involving binary descriptors such as the Binary Robust Independent Elementary Feature (BRIEF) descriptor [23].

For many applications, it is important to derive stable keypoints and keypoint descriptors which are invariant to image scaling and image rotation, and robust with respect to image noise, changes in illumination and small changes in viewpoint.

Satisfying these constraints, SIFT features are commonly applied in a variety of applications which becomes visible in about 9.5 k citations of [53] and about 30.2 k citations of [54], while the use of SURF features has been reported in about 5.6 k citations of [15] and about 6.1 k citations of [16].[1] Both SIFT and SURF features are also typically used in order to detect feature correspondences between intensity images derived for terrestrial laser scans. For each feature correspondence, the respective keypoints may be projected to 3D space by considering the respective range information. This, in turn, yields sparse sets of corresponding 3D points.

4.2.2 Keypoint-Based Point Cloud Registration

Nowadays, most of the approaches for aligning pairs of overlapping scans exploit a keypoint-based point cloud registration which has proven to be among the most efficient strategies. While it has recently been proposed to exploit 3D keypoints [34, 86–88], still most investigations involve forward-projected 2D keypoints detected in image representations of the captured intensity or range information. Hence, we will focus on related work in this research direction.

Once sparse sets of corresponding 3D points have been derived for two scans, the straightforward solution consists of estimating a rigid transformation in the least squares sense [9, 28, 42, 94]. However, the two sparse 3D point sets may also contain some point pairs resulting from incorrect feature correspondences and, consequently, it is advisable to involve the well-known RANSAC algorithm [29] in order to obtain an increased robustness [13, 20, 77].

In case of two coarsely aligned sparse 3D point sets, the well-known Iterative Closest Point (ICP) algorithm [18, 25] and its variants [4, 24, 32, 59, 72] may be applied. The main idea of such an approach is to iteratively minimize a cost function representing the difference between the respective sparse 3D point sets. However, if the coarse alignment between the considered sparse 3D point sets is not good enough, the ICP algorithm may fail to converge or even get stuck in a local minimum instead of the global one. Consequently, such an approach is mainly applied for fine registration.

A further alternative consists of exploiting the spatial information of the derived sparse 3D point sets for a geometric constraint matching based on the 4-Points Congruent Sets (4PCS) algorithm [2] which has recently been presented with the Keypoint-based 4-Points Congruent Sets (K-4PCS) algorithm [86–88]. While the K-4PCS algorithm provides a coarse alignment which is good enough in order to proceed with an ICP-based fine registration, the processing time for the geometric constraint matching significantly increases with the number of 3D points in the sparse 3D point sets due to an evaluation of best matching candidates among point quadruples of both sparse 3D point sets.

---

[1] These numbers were assessed via Google Scholar on May 30, 2015.

4.3   A Novel Framework for Keypoint-Based Point Cloud Registration

Based on our findings from a consideration of related work, we present a novel framework for keypoint-based point cloud registration which is shown in Fig. 4.2, and we explain the involved components addressing the generation of 2D image representations (Sect. 4.3.1), point quality assessment (Sect. 4.3.2), feature extraction and matching (Sect. 4.3.3), the forward projection of 2D keypoints to 3D space (Sect. 4.3.4), correspondence weighting (Sect. 4.3.5), and point cloud registration (Sect. 4.3.6) in the following subsections.

4.3.1   2D Image Representations

In this chapter, we focus on 3D point clouds acquired with terrestrial laser scanners and range cameras. Thereby, terrestrial laser scanners typically capture data by successively considering points on a discrete, regular (typically spherical) raster and recording the respective geometric and radiometric information. In contrast, range cameras have an internal sensor array for simultaneously considering all points on a discrete, regular (typically planar) raster and recording the respective geometric and radiometric information. Thus, for both types of scanning devices, the sampling of data on a discrete, regular raster allows to derive 2D representations for spatial 3D information and the respective radiometric information. In this regard, the spatial 3D information of each scan may be transformed to a range image, where each pixel represents the distance r between the scanning device and the respective 3D scene point, while the captured radiometric information may be transferred to an intensity image. Since the radiometric information depends on both the scene structure and the signal processing unit in the involved scanning device (in terms of signal conversion, amplification, etc.), it is desirable to obtain a normalized image representation which may for instance be obtained by applying a histogram normalization as described in Sect. 2.4 and thus e.g., deriving an 8-bit gray-valued intensity image. Accordingly, we may describe a scan $\mathscr{S} = \{\mathscr{I}_R, \mathscr{I}_I\}$ with respective 2D image representations

Fig. 4.2 The proposed framework for pair-wise point cloud registration: two 3D point clouds serve as input and the output consists of respectively aligned 3D point clouds. Optionally, a forward projection of 2D features to 3D space as well as a weighting of feature correspondences with respect to their suitability may be considered as additional step before point cloud registration

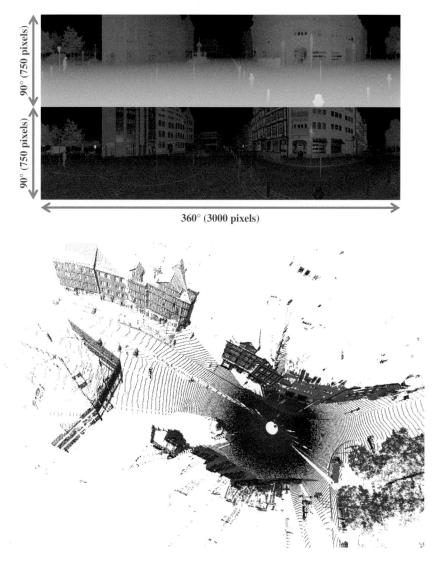

Fig. 4.3 Range image $\mathscr{I}_R$ and intensity image $\mathscr{I}_I$ (*top*), and a visualization of the scan as 3D point cloud where the position of the scanning device is indicated by a *red dot* (*bottom*)

in the form of a range image $\mathscr{I}_R$ and an intensity image $\mathscr{I}_I$ which are visualized in Fig. 4.3 for the example of a terrestrial laser scan acquired with a Riegl LMS-Z360i laser scanner in an urban environment.

4.3.2 Point Quality Assessment

Once the data has been captured, it is worth considering three major issues. First, the used scanning devices are line-of-sight instruments with a certain laser footprint. As a consequence, at edges corresponding to a discontinuity with respect to range information, the laser footprint covers a part of a foreground object and a part of the background which, in turn, results in a "measured" range value between the foreground object and the background. Such erroneous measurements may occur along edges of objects in the scene and have to be removed in order to obtain an appropriate 3D reconstruction of a scene. Second, the range measurements might be corrupted with noise. Whereas the scanned 3D points arising from objects in the scene will probably provide a smooth surface, the scanned 3D points corresponding to the sky tend to be very noisy as they mainly result from atmospheric effects. Third, points on 3D surfaces which are almost perpendicular to the respective line-of-sight (i.e. the incidence angle is relatively small) are more useful for point cloud registration, since the respective range measurements are more accurate. For more detailed considerations, we refer to our investigations in Sect. 2.5, where we present the two measures of range reliability $\sigma_{r,3\times3}$ and local planarity $P_{\lambda,3\times3}$, which allow to quantify the quality of range measurements and may be assigned as attribute to the respective 3D points as indicated in Fig. 4.4 for an exemplary TLS scan.

4.3.3 Feature Extraction and Matching

Once the measured information has been assigned an adequate quality measure, the registration process may rely on both range and intensity information, and a (binary or continuous) confidence map providing the respective quality measure. In order to derive reliable feature correspondences between different scans, we focus on extracting distinctive features from the respectively derived intensity images $\mathscr{I}_1$. For this purpose, local features are well-suited as they allow a reliable matching [49, 93, 99]. Consequently, as explained in Sects. 3.2.5 and 4.2.1.4, we have to address (i) *keypoint detection* in order to adequately retrieve the respective 2D image locations and (ii) *keypoint description* in order to appropriately represent the local image characteristics. Based on the detected keypoints and their respective descriptions, a keypoint matching may be conducted which yields corresponding keypoints between the considered images. The forward projection of the respective 2D image locations of corresponding keypoints to 3D space according to the respective range information yields sparse 3D point clouds of physically (almost) identical 3D points.

   As we do not only consider the intensity information (which would for instance be sufficient for image registration), but also the respective range information for point cloud registration, it is worth reflecting the fundamental ideas of commonly used methods for extracting local features. In particular, the detection of local features has to be taken into account. In general, highly informative image points are located at

Fig. 4.4 Logarithmic representation of the measure $\sigma_{r,3\times3}$ of range reliability and a respective binary confidence map when using a threshold of $t_\sigma = 0.10$ m as well as a representation of the measure $P_{\lambda,3\times3}$ of local planarity and a respective binary confidence map (from *top* to *bottom*)

corners or blobs in an image [10]. Corners in the intensity image often correspond to a discontinuity in the respective range image, and as a consequence, feature detectors which focus on the extraction of image corners (e.g., the Harris corner detector [36]) are less suitable in case of erroneous range measurements due to noise or edge effects. In contrast, image blobs typically correspond to relatively flat regions in the respective range image. Hence, feature detectors such as the Laplacian-of-Gaussian (LoG) detector[2] [50, 51] and its approximations [58] are the methods of choice as well as

---

[2]Note that the Laplacian-of-Gaussian (LoG) filter represents the stable implementation of the Laplacian operator [58].

approaches for finding distinctive image regions such as Maximally Stable Extremal Regions (MSERs) [57]. In this regard, an efficient approximation of the LoG detector is represented by the Difference-of-Gaussian (DoG) detector which is also exploited for the Scale Invariant Feature Transform (SIFT) [53, 54]. The representation of a feature with a descriptor is relatively independent from the applied feature detector, and therefore different feature detector–descriptor combinations may be applied which may generally differ in their suitability, depending on the requirements of the respective application. A recent evaluation for finding the best combination among several feature detectors and various feature descriptors concludes that the DoG/SIFT and MSER detectors combined with a SIFT or DAISY descriptor provide the best performance [26].

Focusing on typical TLS scan representations in the form of panoramic intensity images, the local image characteristics may significantly differ for different scans as shown in Fig. 4.5, particularly for those objects close to the scanning device. Considering the panoramic intensity images derived for different scans along a trajectory with a spacing of approximately 5 m, we may expect both differences in scale (e.g., if we move toward an object) and differences in rotation (e.g., due to the consideration of points on a regular spherical raster). Accordingly, keypoint descriptors may have to cope with significant changes in rotation and scale for changes in the position of the utilized scanning device, and we therefore only involve scale and rotation invariant keypoint representations as listed in Table 4.1. More details on the respective keypoint detectors and descriptors are provided in the following.

4.3.3.1   SIFT

The *Scale Invariant Feature Transform (SIFT)* [53, 54] yields distinctive keypoints at 2D image locations $\mathbf{x} = (x, y)^T \in \mathbb{R}^2$ as well as the respective local descriptors which are invariant to image scaling and image rotation, and robust with respect to image noise, changes in illumination, and reasonable changes in viewpoint. For keypoint detection, the Scale Invariant Feature Transform relies on convolving a given image $\mathcal{I}$ and subsampled versions of $\mathcal{I}$ with Gaussian kernels of variable scale in order to derive the Gaussian scale-space. Subtracting neighboring images in the Gaussian scale-space results in the Difference-of-Gaussian (DoG) pyramid, where extrema in a $(3 \times 3 \times 3)$ neighborhood correspond to keypoint candidates. The 2D locations of these keypoint candidates are refined by an interpolation based on a 3D quadratic function in the scale-space in order to obtain subpixel accurate locations in image space. Furthermore, those keypoint candidates with low contrast which are sensitive to noise as well as those keypoint candidates located along edges which can hardly be distinguished from each other are discarded.

In the next step, each keypoint is assigned its dominant orientation which results for the respective scale from a consideration of the local gradient orientations weighted by the respective magnitude as well as a Gaussian centered at the keypoint. Subsequently, the local gradient information is rotated according to the dominant orientation in order to achieve a rotation invariant keypoint descriptor. The descriptor

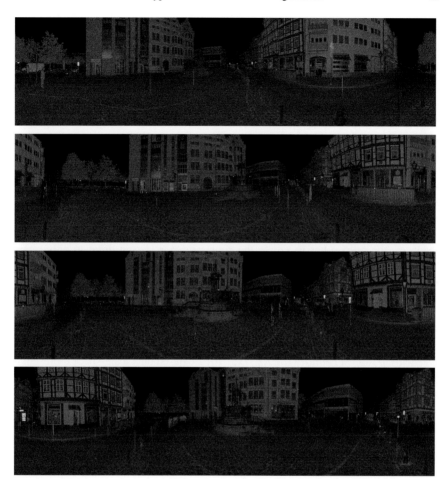

Fig. 4.5 Panoramic intensity images derived for different scans along a trajectory with a spacing of approximately 5 m (from *top* to *bottom*)

Table 4.1 The keypoint detector–descriptor combinations considered in our framework

Variant	Detector	Detector type	Descriptor	Descriptor size/type/information
1	SIFT	Blobs	SIFT	128/float/gradient
2	SURF	Blobs	SURF	64/float/gradient
3	ORB	Corners	ORB	32/binary/intensity
4	A-KAZE	Blobs	M-SURF	64/float/gradient
5	SURF*	Blobs	BinBoost	32/binary/gradient

itself is derived by splitting the local neighborhood into 4×4 subregions. For each of these subregions, an orientation histogram with 8 angular bins is derived by accumulating the gradient orientations weighted by the respective magnitude as well as a Gaussian centered at the keypoint. The concatenation of all histogram bins and a subsequent normalization yield the final 128-dimensional SIFT descriptor.

For deriving feature correspondences, SIFT descriptors are typically compared by considering the ratio ρ of Euclidean distances of a SIFT descriptor belonging to a keypoint in an image to the nearest and second nearest SIFT descriptors belonging to keypoints in another image. A low value of ρ indicates a high similarity to only one of the derived descriptors belonging to the other image, whereas a high value of ρ close to 1 indicates that the nearest and second nearest SIFT descriptors in the other image are quite similar. Consequently, the ratio $\rho \in [0, 1]$ indicates the degree of similarity and thus the distinctiveness of matched features. Since meaningful feature correspondences arise from a greater difference of the considered Euclidean distances, the ratio ρ has to satisfy the constraint $\rho \leq t_{\text{des}}$ in order to only retain meaningful feature correspondences, whereby the threshold t_{des} is typically chosen within the interval $[0.6, 0.8]$.

4.3.3.2   SURF

Speeded-Up Robust Features (SURF) [15, 16] are based on a scale-space representation of the Hessian matrix which is approximated with box filters, so that the elements of the Hessian matrix may efficiently be evaluated at very low computational costs using integral images. Thus, distinctive features in an image correspond to locations in the scale-space where the determinant of the approximated Hessian matrix reaches a maximum in a $(3 \times 3 \times 3)$ neighborhood. The locations of detected maxima are further refined by fitting a 3D quadratic function in the scale-space in order to obtain subpixel accurate locations in image space.

Similar to SIFT, a dominant orientation is calculated for each keypoint. For this purpose, the Haar wavelet responses in x- and y-direction within a circular neighborhood are weighted by a Gaussian centered at the keypoint and represented in a new 2D coordinate frame. Accumulating all responses within a sliding orientation window covering $60°$ yields a local orientation vector, and the orientation vector of maximum length indicates the dominant orientation. In order to obtain a rotation invariant keypoint descriptor, the local gradient information is rotated according to the dominant orientation. Then, the local neighborhood is divided into 4×4 subregions and, for each subregion, the Haar wavelet responses in x'- and y'-direction are weighted by a Gaussian centered at the keypoint. The concatenation of the sum of Haar wavelet responses in x'- and y'-direction as well as the sum of absolute values of the Haar wavelet responses in x'- and y'-direction for all subregions and a subsequent normalization yield the final 64-dimensional SURF descriptor. The comparison of SURF descriptors is carried out in analogy to the comparison of SIFT descriptors.

4.3.3.3 ORB

The approach presented with the *Oriented FAST and Rotated BRIEF (ORB) detector and descriptor* [71] represents a combination of a modified FAST detector and a modified BRIEF descriptor. Consequently, we may have a closer look on the main ideas of the respective basic approaches.

The *Features from Accelerated Segment Test (FAST) detector* [70] analyzes each pixel (x, y) of an image $\mathscr{I}$ and takes into account those pixels located on a surrounding Bresenham circle [22]. The intensity values corresponding to those pixels on the surrounding Bresenham circle are compared to the intensity value $\mathscr{I}(x, y)$. Introducing a threshold t, the investigated pixel (x, y) represents a keypoint candidate if a certain number of contiguous pixels have intensity values above $\mathscr{I}(x, y)+t$ or below $\mathscr{I}(x, y) - t$. A subsequent non-maximum suppression avoids keypoints at adjacent pixels. The modification resulting in the ORB detector is based on employing a scale pyramid of the image, producing FAST features at each level in the pyramid and adding an orientation component to the standard FAST detector.

The *Binary Robust Independent Elementary Feature (BRIEF) descriptor* [23] is derived by computing binary strings from image patches. In this context, the individual bits are obtained from a set of binary tests based on comparing the intensities of pairs of points along specific lines. The modification resulting in the ORB descriptor consists of steering BRIEF according to the orientation of keypoints and thus deriving a rotation-aware version of the standard BRIEF descriptor. The similarity of such binary descriptors may be evaluated using the Hamming distance [35], which is very efficient to compute.

4.3.3.4 A-KAZE and M-SURF

Instead of the Gaussian scale-space of an image, using a nonlinear scale-space may be favorable as Gaussian blurring does not respect the natural boundaries of objects and smoothes to the same degree both details and noise, reducing localization accuracy and distinctiveness of features [7]. Such a nonlinear scale-space may for instance be derived using efficient additive operator splitting (AOS) techniques and variable conductance diffusion which have been employed in order to detect *KAZE features* [7]. The nonlinear diffusion filtering, in turn, makes blurring locally adaptive to the image data and thus reduces noise while retaining object boundaries. However, AOS schemes require solving a large system of linear equations to obtain a solution. In order to increase computational efficiency, it has been proposed to define a nonlinear scale-space with fast explicit diffusion (FED) and thereby embed FED schemes in a pyramidal framework with a fine-to-coarse strategy. Using such a nonlinear scale-space, *Accelerated KAZE (A-KAZE) features* [8] may be extracted by finding maxima of the scale-normalized determinant of the Hessian matrix, where the first- and second-order derivatives are approximated by means of Scharr filters (which approximate rotation invariance significantly better than other popular filters [98]), through the nonlinear scale-space. After a subsequent non-maximum suppression, the remaining keypoint candidates are further refined to subpixel accuracy by fitting a quadratic function to the determinant of the Hessian response in a (3×3) image neighborhood and finding its maximum.

For keypoint description, scale and rotation invariant feature descriptors may be derived by estimating the dominant orientation of the keypoint in analogy to the SURF descriptor and rotating the local image neighborhood accordingly. Based on the rotated neighborhood, using the *Modified-SURF (M-SURF) descriptor* [1] adapted to the nonlinear scale-space has been proposed which, compared to the original SURF descriptor, introduces further improvements due to a better handling of descriptor boundary effects and due to a more robust and intelligent two-stage Gaussian weighting scheme [7].

4.3.3.5   SURF* and BinBoost

Finally, we also involve a keypoint detector–descriptor combination which consists of applying a modified variant of the SURF detector and using the *BinBoost descriptor* [91, 92]. In comparison to the standard SURF detector [15, 16], the modified SURF detector, denoted as SURF* in this work, iterates the parameters of the SURF detector until a desired number of features are obtained.[3] Once appropriate parameters have been derived, the dominant orientation of each keypoint is calculated and used to rotate the local image neighborhood in order to provide the basis for deriving scale and rotation invariant feature descriptors.

As descriptor, the BinBoost descriptor [91, 92] is used which represents a learned low-dimensional, but highly distinctive binary descriptor, where each dimension (each byte) of the descriptor is computed with a binary hash function that was sequentially learned using *boosting*. The weights as well as the spatial pooling configurations of each hash function are learned from training sets consisting of positive and negative gradient maps of image patches. In general, boosting combines a number of weak learners in order to obtain a single strong classifier. In the context of the BinBoost descriptor, the weak learners are represented by gradient-based image features that are directly applied to intensity image patches. During the learning stage of each hash function, the Hamming distance [35] between image patches is optimized, i.e., it is decreased for positive patches and increased for negative patches. Since the BinBoost descriptor with 32 bytes worked best in a variety of experiments, we will only report the results for this descriptor version.

4.3.4   Forward Projection of 2D Keypoints to 3D Space

Since the derived local features are typically localized with subpixel accuracy whereas the measured information is only available for discrete positions on the regular scan grid, the respective geometric and radiometric information as well as

---

[3]This modified version is part of OpenCV 2.4.

the respective quality measures have to be interpolated from the information available for the regular and discrete scan grid, e.g., by applying a bilinear interpolation. Thus, for two considered scans $\mathcal{S}_1$ and $\mathcal{S}_2$, the image locations of corresponding keypoints—which we refer to as 2D/2D feature correspondences $\mathbf{x}_{1,i} \leftrightarrow \mathbf{x}_{2,i}$ in the following—are projected forward to 3D space and used to reduce the captured 3D point cloud data to sparse 3D point clouds of physically (almost) identical 3D points $\mathbf{X}_{1,i} \leftrightarrow \mathbf{X}_{2,i}$ with $\mathbf{X}_{1,i}, \mathbf{X}_{2,i} \in \mathbb{R}^3$. Including the assigned attributes, each feature correspondence may be described with two samples of corresponding information according to

$$s_{1,i} = \left(\mathbf{x}_{1,i}, \mathbf{X}_{1,i}, \sigma_{r,3\times3,1,i}, P_{\lambda,3\times3,1,i}, \rho_{1,i}, \dots\right)$$
$$\leftrightarrow \quad s_{2,i} = \left(\mathbf{x}_{2,i}, \mathbf{X}_{2,i}, \sigma_{r,3\times3,2,i}, P_{\lambda,3\times3,2,i}, \rho_{2,i}, \dots\right) \quad (4.1)$$

where $\sigma_{r,3\times3,1,i}$ and $\sigma_{r,3\times3,2,i}$ as well as $P_{\lambda,3\times3,1,i}$ and $P_{\lambda,3\times3,2,i}$ indicate the quality of the derived 3D points with respect to *range reliability* and *local planarity* (see Sects. 2.5.3 and 2.5.4), and $\rho_{1,i}$ and $\rho_{2,i}$ are the assigned quality measures with respect to the distinctiveness of the used intensity information. Following the commonly applied matching strategy, the process of feature matching is only carried out in one direction and therefore we may define $\rho_i = \rho_{1,i} = \rho_{2,i}$.

As a result from feature matching and the respective forward projection of corresponding 2D keypoints to 3D space, we may exploit either the derived 2D/2D feature correspondences $\mathbf{x}_{1,i} \leftrightarrow \mathbf{x}_{2,i}$ or the derived 3D/3D correspondences $\mathbf{X}_{1,i} \leftrightarrow \mathbf{X}_{2,i}$. However, we may also define 3D/2D correspondences $\mathbf{X}_{1,i} \leftrightarrow \mathbf{x}_{2,i}$. Note that, when exploiting such 3D/2D correspondences and using a terrestrial laser scanner where the image representation is based on a cylindrical or spherical scan grid, the respective 2D image locations $\mathbf{x}_{2,i}$ need to be updated in a subsequent step in a way that they refer to a planar grid.

4.3.5 Correspondence Weighting

The derived correspondences $s_{1,i} \leftrightarrow s_{2,i}$ are typically assigned equal weights and thus treated equally for the registration process, although they may differ in their suitability for the task of point cloud registration. Hence, instead of using equally contributing correspondences, a weight parameter may be derived for each correspondence in order to weight its influence on the estimated transformation. In our experiments, we will only involve binary weights as recently proposed schemes for correspondence weighting need to be further investigated. However, we also want to give an impression about how an appropriate numerical correspondence weighting may be achieved and therefore also present two different approaches in this regard which are particularly interesting in case of using moving sensor platforms equipped with range cameras as scanning devices [101, 106].

Binary Weights As explained in detail in Sect. 2.5, we may derive binary confidence maps distinguishing reliable range measurements from unreliable ones. Accordingly, the forward-projected 2D keypoints may be assigned a respective quality measure such as the measure $\sigma_{r,3\times3}$ of *range reliability* (see Sect. 2.5.3) or the measure $P_{\lambda,3\times3}$ of *local planarity* (see Sect. 2.5.4). Whereas a binary confidence map based on the measure $\sigma_{r,3\times3}$ of range reliability provides almost planar object surfaces for significantly varying incidence angles, the binary confidence map based on the measure $P_{\lambda,3\times3}$ of local planarity only provides object surfaces with almost planar behavior and thereby favors lower incidence angles which tend to yield more accurate range measurements (see Sects. 2.5.5 and 2.5.6).

Our idea consists in discarding the ith feature correspondence with $i = 1, \ldots, N_c$, where N_c represents the number of derived correspondences $s_{1,i} \leftrightarrow s_{2,i}$, if at least one of the respective keypoints corresponds to unreliable range information. Such a removal of feature correspondences, in turn, may be achieved by assigning a weight factor of $w_i = 0$. If the respective keypoints correspond to reliable range information, we assign a weight factor of $w_i = 1$.

Numerical Weights Based on the Theoretical Random Error of Depth Measurements In recent investigations on the registration of RGB-D point cloud data acquired with a Microsoft Kinect [46], it has been proposed to define weight parameters for 3D/3D correspondences based on the theoretical random error of depth measurements. A consideration of the $X^c Y^c$-coordinates in the sensor coordinate frame (where the X^c-axis points to the right and the Y^c-axis points to the bottom) is neglected, since a respective weighting would reduce the contribution of the points with increasing distance from the center of the field-of-view which, in turn, is considered as counterintuitive since off-center points are expected to be more important for an adequate alignment of two RGB-D point clouds.

Numerical Weights Based on Inverse Cumulative Histograms (ICHs) The second approach—which we presented recently [101, 102, 106]—focuses on deriving weight parameters via Inverse Cumulative Histograms (ICHs) which, in turn, rely on our proposed measure $\sigma_{r,3\times3}$ of range reliability presented in Sect. 2.5.3. Considering the generalized definition of a feature correspondence $s_{1,i} \leftrightarrow s_{2,i}$ according to Eq. (4.1) with $\rho_{1,i} = \rho_{2,i} = \rho_i$, a correspondence weighting based on the measure of range reliability exploits the values $\sigma_{r,3\times3,1,i}, \sigma_{r,3\times3,2,i} \in [0\,\mathrm{m}, \infty)$ which are considered as quality measures for the respective 3D points $\mathbf{X}_{1,i}$ and $\mathbf{X}_{2,i}$. More specifically, the influence of the ith feature correspondence $s_{1,i} \leftrightarrow s_{2,i}$ on the registration process is weighted by applying a histogram-based approach, where the histograms are derived via the measure $\sigma_{r,3\times3}$ of range reliability [101, 102]. Thereby, the interval $[0\,\mathrm{m}, 1\mathrm{m}]$ is divided into $N_b = 100$ bins of equal size, and the values $\sigma_{r,3\times3,1,i}$ and $\sigma_{r,3\times3,2,i}$ are mapped to the respective bins $b_{1,j}$ and $b_{2,j}$ with $j = 1, \ldots, N_b$. Those values above the upper boundary of $1\mathrm{m}$ are mapped into the last bin. The occurrence of mappings to the different bins $b_{1,j}$ and $b_{2,j}$ is stored in histograms $\mathbf{h}_1 = [h_{1,j}]_{j=1,\ldots,N_b}$ and $\mathbf{h}_2 = [h_{2,j}]_{j=1,\ldots,N_b}$. From these, the cumulative histograms $\mathbf{h}_{c,1}$ and $\mathbf{h}_{c,2}$ are subsequently derived, where the entries reach from 0 to the number N_c of detected feature correspondences. Since 3D points with lower values $\sigma_{r,3\times3}$ are assumed to be more reliable, they should be assigned a higher weight. For this reason, the *Inverse Cumulative Histograms* (*ICHs*)

$$\mathbf{h}_{c,inv,1} = \left[N_c - \sum_{j=1}^{i} h_{1,j} \right]_{i=1,\dots,N_b} \tag{4.2}$$

and

$$\mathbf{h}_{c,inv,2} = \left[N_c - \sum_{j=1}^{i} h_{2,j} \right]_{i=1,\dots,N_b} \tag{4.3}$$

are derived. Thus, for each feature correspondence $s_{1,i} \leftrightarrow s_{2,i}$, two weight factors are available and we consider the weight factor indicating the least reliable information as significant. Consequently, the respective weight parameter w_i assigned to the ith feature correspondence is finally determined according to

$$w_i = \min \left\{ \mathbf{h}_{c,inv,1}(\sigma_{r,3\times3,1,i}), \mathbf{h}_{c,inv,2}(\sigma_{r,3\times3,2,i}) \right\} \tag{4.4}$$

as the minimum of these values. A positive effect arising from this weighting scheme is that those very unreliable feature correspondences with a value $\sigma_{r,3\times3}$ assigned to the last bin result in a weight parameter which equals 0, and consequently, the respective feature correspondences have no influence on the registration process.

However, we may not only involve the measure $\sigma_{r,3\times3}$ of range reliability in order to weight the influence of the ith feature correspondence, but—particularly when using SIFT and SURF features due to their continuous-valued matching score $\rho_i \in [0, t_{des}]$—also the respective matching score ρ_i encapsulating the reliability of feature matching [106]. Accordingly, we may also divide the interval $[0, t_{des}]$ into $N_b = 100$ bins of equal size, map the values ρ_i to the respective bins b_j, and derive a histogram $\mathbf{h} = [h_j]_{j=1,\dots,N_b}$. Based on this histogram, we may further derive the cumulative histogram $\mathbf{h}_c$ as well as an inverse cumulative histogram $\mathbf{h}_{c,inv}$ with

$$\mathbf{h}_{c,inv} = \left[N_c - \sum_{j=1}^{i} h_j \right]_{i=1,\dots,N_b} \tag{4.5}$$

which provides an additional weight factor for each correspondence $s_{1,i} \leftrightarrow s_{2,i}$. Thus, a weighting scheme taking into account all these derived weight factors $\mathbf{h}_{c,inv,1}(\sigma_{r,3\times3,1,i})$, $\mathbf{h}_{c,inv,2}(\sigma_{r,3\times3,2,i})$ and $\mathbf{h}_{c,inv}(\rho_i)$ for each feature correspondence $s_{1,i} \leftrightarrow s_{2,i}$ may be defined by considering the weight factor indicating the least reliable information as significant and therefore determining the respective weight parameter w_i assigned to the ith feature correspondence according to

$$w_i = \min \left\{ \mathbf{h}_{c,inv,1}(\sigma_{r,3\times3,1,i}), \mathbf{h}_{c,inv,2}(\sigma_{r,3\times3,2,i}), \mathbf{h}_{c,inv}(\rho_i) \right\} \tag{4.6}$$

as the minimum of these values.

In order to avoid the manual selection of parameters such as thresholds or the number of histogram bins, this weighting scheme may further be modified from a discrete description based on histograms to a continuous description involving probability density functions. Note that the described histograms represent the empirical distribution of the derived quality values $\sigma_{r,3\times3} \in [0\,\text{m}, 1\,\text{m}]$ and $\rho \in [0, t_{\text{des}}]$. A normalization of each entry by the total number of samples yields the normalized histogram which represents an empirical density function. The resulting normalized cumulative histogram thus represents an empirical cumulative density function. Hence, the transfer to a continuous description only requires the calculation of a continuous probability density function. The main idea to achieve such a continuous description is to treat each quality measure independently as a continuous random variable X. Such a random variable X has a probability density function p, where p is a nonnegative Lebesgue-integrable function, if $P\,(a \leq X \leq b) = \int_a^b p\,(\xi)\,d\xi$ is satisfied on the interval $[a, b]$. The probability that X will be less than or equal to τ may thus be described with

$$F\,(\tau) = P\,(X \leq \tau) = \int_{-\infty}^{\tau} p\,(\xi)\,d\xi \qquad (4.7)$$

where $F\,(\tau)$ represents a monotonically nondecreasing right-continuous function with $\lim_{\tau \to -\infty} F\,(\tau) = 0$ and $\lim_{\tau \to +\infty} F\,(\tau) = 1$. In order to satisfy these constraints, a nonparametric probability density function $p\,(\tau)$ may be fitted to the calculated values by exploiting a normal kernel $K_N\,(\tau)$ with

$$K_N\,(\tau) = \frac{1}{\sqrt{2\pi}} \exp\left\{-\frac{\|\tau\|^2}{2}\right\}. \qquad (4.8)$$

This kernel satisfies $\int_{\mathbb{R}} K_N\,(\xi)\,d\xi = 1$ and represents a bounded function with compact support. Subsequently, the *cumulative distribution function (CDF)* denoted as $P\,(\tau)$ with

$$P\,(\tau) = \int_0^{\infty} p\,(\xi)\,d\xi \qquad (4.9)$$

may be derived. The upper tail of this cumulative distribution function also known as *upper cumulative distribution function* and formally described with

$$P\,(X > \tau) = 1 - P\,(X \leq \tau) \qquad (4.10)$$
$$= 1 - \text{CDF}\,(\tau) \qquad (4.11)$$
$$= \text{ICDF}\,(\tau) \qquad (4.12)$$

represents the probability that X is above the given value τ. If this probability is high, the current value τ is considered to be reliable. Accordingly, we may follow the definition of the Inverse Cumulative Histogram (ICH) and define the upper cumulative distribution function as *Inverse Cumulative Distribution Function*[4] (*ICDF*). As a result, the weight of the *i*th feature correspondence $s_{1,i} \leftrightarrow s_{2,i}$ may be derived as

$$w_i = \min \left\{ \text{ICDF}_1(\sigma_{r,3\times3,1,i}), \text{ICDF}_2(\sigma_{r,3\times3,2,i}), \text{ICDF}(\rho_i) \right\}. \tag{4.13}$$

4.3.6 Point Cloud Registration

By conducting a forward projection of n corresponding 2D keypoints $\mathbf{x}_{1,i} \leftrightarrow \mathbf{x}_{2,i}$ with $i = 1, \ldots, n$ between the intensity images of two scans $\mathscr{S}_1$ and $\mathscr{S}_2$ according to the respective range information, sparse sets of corresponding 3D points $\mathbf{X}_{1,i} \leftrightarrow \mathbf{X}_{2,i}$ have been derived. Classically, the task of keypoint-based point cloud registration is solved by estimating the rigid Euclidean transformation between the two sets of corresponding 3D points, i.e., a *rigid transformation* of the form

$$\mathbf{X}_{2,i} \approx \hat{\mathbf{X}}_{2,i} = \mathbf{R}\mathbf{X}_{1,i} + \mathbf{t} \tag{4.14}$$

with a rotation matrix $\mathbf{R} \in \mathbb{R}^{3\times3}$ and a translation vector $\mathbf{t} \in \mathbb{R}^3$. Accordingly, the rigid transformation is estimated in object space.

As feature correspondences between image representations in the form of panoramic range and intensity images are available, we may not only be interested in considering 3D cues in terms of how good corresponding 3D points fit together with respect to the Euclidean distance, but also in simultaneously considering 2D cues in terms of how good the 2D locations in the image representations fit together. As a result of taking into account both 3D and 2D cues, an improvement in accuracy may be expected for the registration results. Accordingly, we propose to estimate the transformation between considered 3D point clouds in observation space by transferring the task of point cloud registration to (i) the task of solving the P*n*P problem and (ii) the task of solving the relative orientation problem.

4.3.6.1 Solving the PnP Problem—Projective Scan Matching

Our first strategy has been inspired by the idea of deriving synthetic camera images by projecting scans with the associated intensity information onto the image planes of virtual cameras and minimizing discrepancies in color, range, and silhouette between pairs of images [66]. While this approach is rather impractical for large 3D point clouds and thus for the registration of TLS scans, we propose a keypoint-based projective scan matching strategy [100, 103, 104], where only the forward-projected

---

[4]Note that this definition does not correspond to calculating the inverse of a function which would be another function taking the output of the initial function as its input and reproducing the input of the initial function as its output.

2D keypoints are back-projected onto the image plane of a virtual pinhole camera in order to derive 3D/2D correspondences which, in turn, allow to transfer the task of point cloud registration to the task of solving the well-known *Perspective-n-Point* *(PnP) problem*, where the aim is to estimate the exterior orientation or pose of a camera from a set of n correspondences between 3D points $\mathbf{X}_i$ of a scene and their 2D projections $\mathbf{x}_i$ in the image plane of a camera [29, 108]. Thereby, the derived 3D/2D correspondences $\mathbf{X}_i \leftrightarrow \mathbf{x}_i$ serve as input for an efficient scheme taking into account both 3D cues based on the considered 3D point clouds and 2D cues based on the derived 2D imagery, and thus highly accurate registration results may already be expected for coarse registration.

Coarse Registration In order to transfer the task of point cloud registration to the task of solving the PnP problem, respective 2D observations are required on a 2D image plane. This constraint is satisfied when using modern range cameras such as Microsoft Kinect, PMD[vision] CamCube 2.0, or MESA Imaging SR4000. When using terrestrial laser scanners, however, the derived 2D image representations are based on a regular (either cylindrical or spherical) scan grid where the measurements are captured with a certain angular resolution in horizontal and vertical directions. As a consequence, virtual 2D projections have to be introduced via a projective projection, e.g., in the form of synthetic camera images which may easily be generated from the intensity values via back-projection onto a regular 2D grid using a standard pinhole camera model [30]. The rigid alignment of a virtual camera to each scan may also be used in order to project a surface from one scan onto the image plane of the virtual camera belonging to a different scan [66] which, in turn, allows a comparison with respect to range, color, and silhouette properties. Instead of back-projecting all measured 3D points onto the image plane of a virtual camera, we only back-project those 3D points of the sparse 3D point sets derived via 2D feature extraction and a respective forward projection to 3D space based on the respective range information in order to increase computational efficiency.

Without loss of generality, we may assume that—when considering a scan pair $\mathscr{P} = \{\mathscr{S}_1, \mathscr{S}_2\}$ as illustrated in Fig. 4.6—the position and orientation of scan $\mathscr{S}_1$ is known with respect to the world coordinate frame. Consequently, for scan $\mathscr{S}_1$, the respective forward projection of 2D keypoints $\mathbf{x}_{1,i} \in \mathbb{R}^2$ results in 3D coordinates $\mathbf{X}_{1,i} \in \mathbb{R}^3$ which are also known with respect to the world coordinate frame, whereas the forward projection of 2D keypoints $\mathbf{x}_{2,i} \in \mathbb{R}^2$ for scan $\mathscr{S}_2$ results in 3D coordinates $\mathbf{X}_{2,i} \in \mathbb{R}^3$ which are only known with respect to the local coordinate frame of the scanning device. The basic idea of projective scan matching consists of introducing 2D cues by back-projecting the 3D points $\mathbf{X}_{2,i}$ onto a virtual image plane for which the projection model of a pinhole camera is exploited [104]:

$$\mathbf{x}_{2,i}^* = \mathbf{K}_C \; [\mathbf{R}_C | \mathbf{t}_C] \; \mathbf{X}_{2,i} \tag{4.15}$$

In this equation, the calibration matrix of the virtual camera is denoted with $\mathbf{K}_C$ and arbitrary parameters may be selected in order to specify the focal lengths and the principal point [100, 104]. Note that the image plane may even not necessarily have to

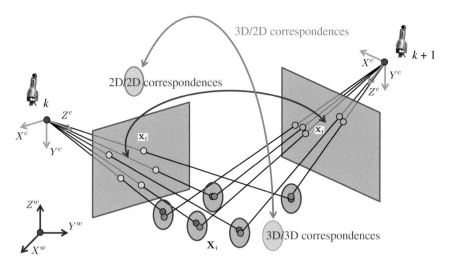

Fig. 4.6 The principle of deriving 3D/2D correspondences for two considered scans $\mathscr{S}_k$ and $\mathscr{S}_{k+1}$

be limited on a finite area as we are only interested in the respective 2D coordinates. Furthermore, the rotation matrix $\mathbf{R}_C \in \mathbb{R}^{3 \times 3}$ describes the relative orientation of the virtual camera with respect to the local coordinate frame of the laser scanner and the translation vector $\mathbf{t}_C \in \mathbb{R}^3$ represents the respective translation between the origin of the local coordinate frame of the laser scanner and the projective center of the virtual camera. Without loss of generality, we may assume that the relative orientation of the virtual camera remains fixed with respect to the local coordinate frame of the laser scanner for the considered scans. Assuming such a rigid alignment of the virtual camera with respect to the local coordinate frame of the laser scanner, the rotation matrix $\mathbf{R}_C$ and the translation vector $\mathbf{t}_C$ remain constant across all considered scans. Involving a virtual camera in this manner, we may specify both its interior behavior and its pose arbitrarily without loss of generality. Accordingly, we specify them in a way that the virtual camera looks into the horizontal direction and that the projective center of the virtual camera coincides with the location of the laser scanner, i.e., $\mathbf{t}_C = \mathbf{0}$. The points $\mathbf{x}_{2,i}^* \in \mathbb{R}^2$, in turn, allow to transform the n 2D/2D feature correspondences $\mathbf{x}_{1,i} \leftrightarrow \mathbf{x}_{2,i}$ with $i = 1, \dots, n$ to 3D/2D correspondences $\mathbf{X}_{1,i} \leftrightarrow \mathbf{x}_{2,i}^*$ which are required in order to solve the PnP problem.

Due to the introduced 2D projections, we refer the resulting approaches to as *Projective Scan Matching (PSM)* techniques. In order to solve the PnP problem, a variety of approaches may be applied. A robust and efficient approach for solving the PnP problem has been proposed with the *Efficient Perspective-n-Point (EPnP) algorithm* [48, 60] which represents a noniterative method and provides an accurate solution to the PnP problem with only linear complexity. Compared to other approaches for solving the PnP problem, this algorithm is not only fast and accurate, but also designed to work with a large number of correspondences and it does not require an initial estimate. The EPnP algorithm is based on the idea of expressing the n 3D scene points $\mathbf{X}_i$ whose coordinates are known in the world coordinate frame as a weighted sum of four virtual and noncoplanar control points $\mathbf{C}_j$ for general configurations according to

$$\mathbf{X}_i = \sum_{j=1}^{4} \alpha_{ij}\, \mathbf{C}_j \qquad (4.16)$$

with

$$\sum_{j=1}^{4} \alpha_{ij} = 1 \qquad (4.17)$$

where the parameters α_{ij} represent homogeneous barycentric coordinates. As, in the world coordinate frame indicated with a superscript ^{w}, the points $\mathbf{X}_i^w$ are known as well as the control points $\mathbf{C}_j^w$, the parameters α_{ij} are known for the same equation in camera coordinates indicated with a superscript ^{c}. Hence, the points $\mathbf{X}_i^c$ may be expressed via the control points $\mathbf{C}_j^c$ and thus each 3D/2D feature correspondence provides a relation of the form

$$w_i \begin{bmatrix} \mathbf{x}_i \\ 1 \end{bmatrix} = \mathbf{K}_C\, \mathbf{X}_i^c = \mathbf{K}_C \sum_{j=1}^{4} \alpha_{ij}\, \mathbf{C}_j^c \qquad (4.18)$$

for $i = 1, \ldots, n$, where w_i are scalar projective parameters and $\mathbf{K}_C$ describes the calibration matrix, in our case the one of the virtual camera. Expanding the previous equation yields a linear system consisting of three equations given by

$$w_i \begin{bmatrix} x_i \\ y_i \\ 1 \end{bmatrix} = \begin{bmatrix} f_x & 0 & x_0 \\ 0 & f_y & y_0 \\ 0 & 0 & 1 \end{bmatrix} \sum_{j=1}^{4} \alpha_{ij} \begin{bmatrix} X_j^c \\ Y_j^c \\ Z_j^c \end{bmatrix} \qquad (4.19)$$

with the focal length coefficients f_x and f_y, the coordinates x_0 and y_0 of the principal point, and the coordinates $\begin{bmatrix} X_j^c, Y_j^c, Z_j^c \end{bmatrix}^T$ of the virtual control points $\mathbf{C}_j$. This linear system has 12 unknown parameters from the control points and, additionally, n unknown parameters w_i. As can be seen from the third equation, the scalar projective parameters w_i may be determined according to

$$w_i = \sum_{j=1}^{4} \alpha_{ij}\, Z_j^c \qquad (4.20)$$

and substituted into the other two equations, respectively. Concatenating the two resulting modified equations for all n 3D/2D correspondences yields a linear equation system

$$\mathbf{M}\,\mathbf{y} = \mathbf{0} \qquad (4.21)$$

with a matrix $\mathbf{M} \in \mathbb{R}^{2n \times 12}$ and a vector $\mathbf{y} \in \mathbb{R}^{12}$ containing the 3D coordinates of the four virtual control points $\mathbf{C}_j$. Thus, only the coordinates of the control points in the camera coordinate system have to be estimated as they are directly related with the coordinates $\mathbf{X}_i^c$ of all 3D points which is the main reason for the efficiency of the EPnP algorithm. Once both the world coordinates and the camera coordinates of the 3D points $\mathbf{X}_i$ are known, the transformation parameters aligning both coordinate frames may be retrieved via standard methods involving a closed-form solution in the least squares sense [9, 42]. For more details as well as a further optimization step relying on the Gauss–Newton algorithm, we refer to [48, 60].

For a robust estimation in case of existing outlier correspondences, the RANSAC algorithm [29] represents the method of choice as it eliminates the influence of outlier correspondences which are not in accordance with the largest consensus set supporting the given transformation model. Following the original implementation [60], the RANSAC-based EPnP scheme relies on selecting small, but not minimal subsets of seven correspondences in order to estimate the model parameters and check the whole set of correspondences for consistent samples. In comparison to minimal subsets, this further reduces the sensitivity to noise. In order to avoid testing all possible subsets, which would be very time-consuming, we exploit an efficient variant, where the number of iterations—which equals the number of randomly chosen subsets— is selected high enough, so that a subset including only inlier correspondences is selected with a certain probability [29, 37].

Outlier Removal and Fine Registration When involving approaches for extracting and matching local features, we might face a certain number of erroneous feature correspondences due to the high visual similarity of the respective local image neighborhoods. In urban environments, for instance, periodic shapes of façades are likely to occur and—due to the representation of intensity information in the form of panoramic intensity images—a feature detected in the intensity image derived for a scan $\mathscr{S}_1$ may be very similar to a different feature detected in the intensity image derived for another scan $\mathscr{S}_2$, while the correct assignment is not possible because of the distortions in the respective intensity image derived for scan $\mathscr{S}_2$. Thus, it is advisable to check if all detected feature correspondences are really suitable with respect to the corresponding 3D points and remove those feature correspondences indicating potential outlier correspondences from the sparse 3D point clouds. This may be achieved by introducing geometric constraints based on the 3D distances between the coarsely aligned pairs of corresponding 3D points. More specifically, the sparse 3D point set $\mathbf{X}_{1,i}$ is transformed to $\hat{\mathbf{X}}_{2,i}$ using Eq. (4.14) and the coarse estimates for $\mathbf{R}$ and $\mathbf{t}$. Subsequently, the Euclidean distance between the coarsely aligned pairs of corresponding 3D points is calculated and large distances indicate erroneous correspondences which have to be discarded. For the latter, a respective criterion may be based on the distribution of the distance between the coarsely aligned pairs of corresponding 3D points. By sorting the corresponding 3D points according to their 3D distances after the coarse alignment, we may consider a set of correspondences representing the third of the smallest distances, and iteratively adding the correspondence with the next greater distance allows to check whether the new sample takes a strong influence on the mean value of the set which is done by considering the first derivative of

$$\tilde{d}(i) = \frac{\frac{1}{i}\sum_{j=1}^{i} d_j}{\frac{1}{N_c}\sum_{j=1}^{N_c} d_j}.$$

(4.22)

Alternatively, we could also only consider those points with, e.g., a distance below 1 m. In order to remove such heuristics, one could employ iterative reweighted least squares techniques or a RANSAC-based modification of the ICP algorithm.

Finally, the remaining 3D points of the sparse 3D point sets are provided as input for (i) a refinement via a repetition of the RANSAC-based EPnP scheme [104], (ii) a standard fine registration based on the ICP algorithm [100], or (iii) a fine registration based on the estimation of a standard rigid transformation in the least squares sense [102]. A general processing workflow of such projective scan matching techniques is visualized in Fig. 4.7. Note that, in this regard, the ICP algorithm generally converges to the nearest local minimum of a mean square distance metric, where the rate of convergence is high for the first few iterations. Given an appropriate coarse registration delivering the required initial values for $\mathbf{R}$ and $\mathbf{t}$, even a global minimization may be expected. In our experiments, we apply an ICP-based fine registration and consider the result after 10 iterations.

4.3.6.2 Solving the Relative Orientation Problem—Omnidirectional Scan Matching

As omnidirectional representations in the form of panoramic range and intensity images are available, our second strategy presented in [95] consists of estimating the transformation in observation space, i.e., we intend to find the relative orientation between consecutive scans directly. For this purpose, we exploit the *bearing vectors* defined by the origin of the local coordinate frame and the forward-projected 2D keypoints. The main idea behind such an approach is based on the fact that 2D keypoints may be localized up to subpixel accuracy, whereas the range measurement may not always be that reliable (see Sect. 2.5). As a consequence, we may assume that bearing vectors may be determined with a higher reliability in comparison to range measurements.

More specifically, we apply a spherical normalization $\mathsf{N}(\cdot)$ which normalizes forward-projected 2D keypoints $\mathbf{X}_{j,i}$ given in the local coordinate frame of scan $\mathscr{S}_j$ with $j = \{1, 2\}$ to unit length and thus yields the bearing vectors

$$\mathbf{v}_{j,i} = \mathsf{N}(\mathbf{X}_{j,i}) = \frac{\mathbf{X}_{j,i}}{\|\mathbf{X}_{j,i}\|}$$

(4.23)

that simply represent the direction of a 3D point $\mathbf{X}_{j,i}$ with respect to the local coordinate frame of the laser scanner. Thus, the task of point cloud registration may be transferred to the task of finding the transformation of one set of bearing vectors to

Fig. 4.7 The principle of
Projective Scan Matching
(PSM) for two considered
scans $\mathcal{S}_k$ and $\mathcal{S}_{k+1}$: the
different alternatives for a
refinement of the coarse
estimate obtained via a
RANSAC-based EPnP
scheme are indicated for (*1*)
a repetition of the
RANSAC-based EPnP
scheme, (*2*) ICP-based fine
registration, and (*3*) fine
registration based on
estimating a standard rigid
transformation (SRT) in the
least squares sense

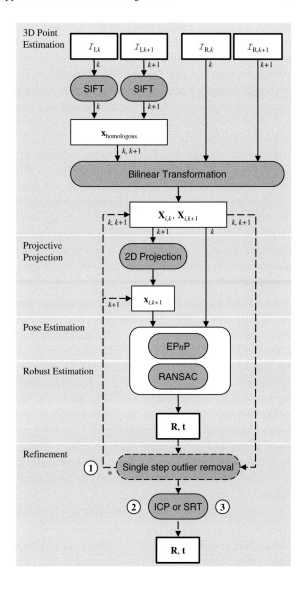

another. In photogrammetry and computer vision, this is known as *relative orientation problem* and the transformation is encoded both in the essential matrix **E** and
the fundamental matrix **F**, respectively. The relationship between both is given by

$$\mathbf{F} = \mathbf{K}^{-T}\mathbf{E}\mathbf{K}^{-1} = \mathbf{K}^{-T}[\mathbf{t}]_\times \mathbf{R}\mathbf{K}^{-1} \qquad (4.24)$$

where **K** is a calibration matrix, **R** represents a rotation matrix, and $[\mathbf{t}]_\times$ denotes the
skew symmetric matrix of the translation vector **t**. For omnidirectional or panoramic

images, the standard fundamental matrix $\mathbf{F}$ cannot be estimated, since it encapsulates the calibration matrix $\mathbf{K}$ which, in turn, is based on perspective constraints that do not hold in the omnidirectional case. The essential matrix $\mathbf{E}$, however, is independent of the camera type and may hence be used to estimate the transformation between two panoramic images. Focusing on the consideration of panoramic range and intensity images, our strategy addresses the omnidirectional case—thus it may also be referred to as *omnidirectional scan matching (OSM)*—and therefore relies on the calculation of the essential matrix $\mathbf{E}$.

In general, at least five 2D/2D correspondences are required in order to calculate a finite number of solutions for the essential matrix $\mathbf{E}$ [64] and various algorithms for estimating $\mathbf{E}$ exist. These algorithms include the minimal five-point algorithm [61, 83], the six-point algorithm [65], the seven-point algorithm [37], and the eight-point algorithm [52]. As observed in seminal work [68, 83] and also verified by own tests, the five-point solver presented in [83] performs best in terms of numerical precision, special motions or scene characteristics, and resilience to measurement noise. For this reason, we only involve this algorithm of the category addressing omnidirectional scan matching.

Coarse Registration The input for our registration procedure is represented by putative 2D/2D feature correspondences between 2D points $\mathbf{x}_{j,i}$ that are subsequently transformed to bearing vectors $\mathbf{v}_{j,i}$ by exploiting the respective range information. Since the 2D points $\mathbf{x}_{j,i}$ are localized with subpixel accuracy, a respective bilinear interpolation is applied on the 2D scan grid in order to obtain the respective 3D coordinates $\mathbf{X}_{j,i}$. Then, the essential matrix $\mathbf{E}$ is estimated using the five-point algorithm presented in [83] and thereby involving the RANSAC algorithm [29] for increased robustness. For this, we use the implementation of OpenGV [47]. A subsequent decomposition of the essential matrix $\mathbf{E}$ yields the rotation matrix $\mathbf{R}$ and the translation vector $\hat{\mathbf{t}}$ [37]. Since the essential matrix $\mathbf{E}$ only has five degrees of freedom, the translation vector $\hat{\mathbf{t}}$ is only known up to a scalar factor s, which is indicated by a $\hat{\;}$ symbol. In order to recover the scale factor s, the following is calculated over all inliers:

$$s_{\text{median}} = \underset{i}{\text{median}} \left(\| \mathbf{X}_{2,i} - \mathbf{R}\mathbf{X}_{1,i} \| \right) \tag{4.25}$$

The median is used to diminish potential outliers that could still reside in the data. Finally, the direction vector $\hat{\mathbf{t}}$ is scaled by s_{median} in order to derive the translation vector $\mathbf{t} = s_{\text{median}}\hat{\mathbf{t}}$.

Outlier Removal and Fine Registration In order to remove those 3D points indicating potential outlier correspondences from the sparse 3D point clouds, we apply a simple heuristic. First, the point set $\mathbf{X}_{1,i}$ is transformed to $\hat{\mathbf{X}}_{2,i}$ using Eq. (4.14) and the coarse estimates for $\mathbf{R}$ and $\mathbf{t}$. Then, the Euclidean distance between all corresponding 3D points is calculated and only those point pairs with an Euclidean distance below 1 m are kept. In order to remove such heuristics, one could employ iterative reweighted least squares techniques or a RANSAC-based modification of the ICP algorithm. The remaining 3D points of the sparse 3D point sets are provided as input for a standard ICP algorithm [18] which generally converges to the nearest local minimum of a mean square distance metric, where the rate of convergence is high for the first few iterations. Given an appropriate coarse registration delivering the required initial values for **R** and **t**, even a global minimization may be expected. In our experiments, we apply an ICP-based fine registration and consider the result after 10 iterations.

4.4 Experimental Results

In order to test the performance of our framework for point cloud registration, we involve a publicly available benchmark dataset (Sect. 4.4.1) and conduct experiments focusing on different aspects (Sect. 4.4.2). The derived results are presented in the form of an extensive evaluation (Sect. 4.4.3), where we address the different aspects.

4.4.1 Dataset

In order to facilitate an objective comparison between the results of our approaches and those results of other recent or future methodologies, we evaluate the performance of our framework on a publicly available benchmark TLS dataset. The involved TLS dataset[5] has been acquired with a Riegl LMS-Z360i laser scanner in an area called "Holzmarkt" which is located in the historic district of Hannover, Germany. According to [21], the Riegl LMS-Z360i has a single-shot measurement accuracy of 12 mm and its field-of-view covers $360° \times 90°$, while the measurement range reaches up to 200 m. Furthermore, the angular resolution is about $0.12°$ and thus a full scan results in $3000 \times 750 = 2.25$ M scanned 3D points. Since both range and intensity information are available for each point on the discrete scan grid, 2D representations in the form of panoramic range and intensity images may easily be derived (Fig. 4.3).

In total, the dataset consists of 20 scans of which 12 were taken with (approximately) upright scan head and 8 with a tilted scan head. The single scan positions for the upright scans have been selected systematically along a trajectory with a spacing of approximately 5 m, whereas the scan positions for the tilted scans almost coincide with the scan position for an upright scan, and reference values for both position and orientation have been obtained by placing artificial markers in the form of retro-reflective cylinders in the scene and carrying out a manual alignment based on these artificial targets. Thus, errors in the range of a few millimeters may be expected. In our experiments, we consider the similarity between upright and tilted scans acquired at almost the same position as too high to allow a fair statement on the registration accuracy obtained with our framework (since the respective errors with respect to the estimated scan position are significantly below the measurement accuracy of 12 mm),

---

[5]This benchmark TLS dataset and other datasets have been released at http://www.ikg.uni-hannover. de/index.php?id=413\&L=de (last access: 30 May 2015).

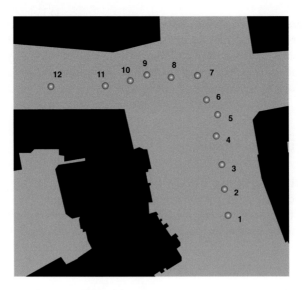

Fig. 4.8 Map of the Hannover "Holzmarkt": the position of buildings is visualized in *dark gray* and the scan positions for different scans $\mathscr{S}_j$ are indicated with *red spots*. The scan IDs are adapted according to [21]

and hence we only use the upright scans (Fig. 4.8). The respective reference values for the relative orientation between different scan pairs are provided in Table 4.2.

Since both range and intensity information are recorded for each point on the discrete scan raster, we may easily characterize each scan with a respective panoramic range image $\mathscr{I}_R$ and a respective panoramic intensity image $\mathscr{I}_I$, where each image has a size of 3000×750 pixels. As the captured intensity information depends on the device, we adapt it via histogram normalization to the interval $[0, 255]$ in order to obtain an 8-bit gray-valued image $\mathscr{I}_I$.

4.4.2   Experiments

Our experiments focus on the successive pair-wise registration of scan pairs $\mathscr{P}_j = \{\mathscr{S}_j, \mathscr{S}_{j+1}\}$ with $j = 1, \ldots, 11$. In order to evaluate the performance of our framework, we consider the respective position and angle errors after coarse and fine registration. Thereby, the position errors indicate the deviation of the estimated scan positions from the reference positions, whereas the angle errors indicate the deviation of the estimated angles from the reference angles. Furthermore, we transform the estimated rotation matrix and the respective reference to Rodrigues vectors which, in turn, allow to derive angle errors as the difference of these Rodrigues vectors with respect to their length [47].

Table 4.2 Reference values for the relative orientation and overlap of pairs $\mathscr{P}_j = \{\mathscr{S}_j, \mathscr{S}_{j+1}\}$ of successive scans $\mathscr{S}_j$ and $\mathscr{S}_{j+1}$ according to [21] as well as the distance $d_{j,j+1}$ between the respective positions of the laser scanner

Scan pair $\mathscr{P}_j$	X (m)	Y (m)	Z (m)	ω (°)	ϕ (°)	κ (°)	Overlap (%)	$d_{j,j+1}$ (m)
$\mathscr{P}_1$	−5.50	0.96	0.02	−1.088	−0.112	51.731	83.1	5.58
$\mathscr{P}_2$	−2.50	4.64	0.08	1.432	−0.958	5.733	82.6	5.27
$\mathscr{P}_3$	−2.72	5.47	0.01	0.824	−1.174	61.834	81.3	6.11
$\mathscr{P}_4$	3.58	2.90	−0.08	1.482	2.238	122.148	83.6	4.61
$\mathscr{P}_5$	3.07	−2.51	0.07	0.096	0.665	147.948	80.3	3.97
$\mathscr{P}_6$	−5.01	2.50	−0.01	−0.701	0.150	149.118	81.0	5.60
$\mathscr{P}_7$	1.18	5.69	0.07	−1.661	−0.525	−11.598	81.3	5.81
$\mathscr{P}_8$	0.73	5.13	0.14	0.664	0.829	−135.342	81.0	5.18
$\mathscr{P}_9$	−2.04	−3.36	0.04	−0.115	1.954	−137.813	82.9	3.93
$\mathscr{P}_{10}$	5.34	0.82	−0.19	1.530	1.731	48.711	77.2	5.41
$\mathscr{P}_{11}$	7.92	−8.84	0.08	0.582	0.292	−132.744	74.9	11.87

Consequently, we first address the organization of unordered scans with respect to their similarity in order to provide the basis for a successive pair-wise registration of TLS scans in Sect. 4.4.3.1. Based on the derived scan order, we subsequently focus on our projective scan matching technique via the RANSAC-based EPnP scheme in Sect. 4.4.3.2 and conduct a detailed evaluation on a benchmark TLS dataset. This is followed by a deeper investigation of the impact of a removal of unreliable feature correspondences by exploiting different point quality measures in Sect. 4.4.3.3. Finally, in Sect. 4.4.3.4, we focus on omnidirectional scan matching and thereby also evaluate the impact of different feature detector–descriptor combinations.

4.4.3 Results

In the following, we present the results derived by applying our framework and thereby focus on different aspects in respective subsections.

4.4.3.1 Scan Organization for a Successive Pair-Wise Registration

In an initial step, it might be desirable to derive a measure describing the similarity of different scans which may be exploited in order to automatically organize a given set of unorganized scans for a successive pair-wise registration. This is particularly important for approaches relying on the use of features extracted from the respective range and intensity images, since a higher overlap of considered scans results in a higher similarity and thus more feature correspondences which, in turn, increases the robustness of respective registration approaches. In this regard, it has been proposed to derive a topological graph, where the nodes represent the single scans and the edges describe their similarity, e.g., based on the number of matched lines determined from the range information [79]. Inspired by this approach, we propose to exploit the number of feature correspondences between respective intensity images [100]. The smaller the weight of an edge is, the smaller is the overlap between the scans corresponding to the nodes connected by this edge. Thus, an appropriate scan order for a reliable successive pair-wise registration may be derived via a *minimum spanning tree* [44].

More specifically, we take into account that any set of unorganized scans may directly be represented as a graph, where the nodes represent the scans and the edges are weighted with the total number of feature correspondences between the intensity images derived for the respective scans. In the most general case, every node is connected with every other node which results in a complete graph, and the number of feature correspondences between all possible scan pairs allows to derive the confusion matrix $\mathbf{C}_S$ indicating the similarity of respective scans. The diagonal elements $\mathbf{C}_S(j, j)$ for $j = 1, \ldots, 12$ represent the total number of features extracted in the respective intensity images $\mathscr{I}_{1,j}$. The derived confusion matrix $\mathbf{C}_S$ when using the same scan IDs as in the given TLS dataset and SIFT features in order to derive feature correspondences ($t_{\text{des}} = 0.66$) is provided in Table 4.3. This table reveals that the confusion matrix is not necessarily symmetric which is due to the ratio ρ of Euclidean distances of a SIFT descriptor belonging to a keypoint in an image to the nearest and second nearest SIFT descriptors in another image. For instance, if

Table 4.3 Confusion matrix $\mathbf{C}_S$ indicating the number of feature correspondences between the intensity images of different scans within the chosen TLS dataset

ID	01	02	03	05	06	08	09	11	13	15	17	19
01	4986	217	63	45	33	58	41	28	23	44	62	39
02	229	5663	319	100	59	80	43	48	24	35	46	38
03	88	308	5967	253	120	56	47	68	30	57	38	56
05	70	114	277	6200	484	78	58	68	25	131	68	84
06	31	70	124	466	6682	169	68	56	21	477	134	71
08	86	96	53	78	163	6867	205	64	31	328	404	99
09	39	34	37	56	68	158	5571	330	22	78	577	656
11	17	24	37	40	44	41	277	4061	77	24	134	408
13	8	9	6	8	20	12	25	62	3308	7	29	30
15	61	40	59	129	503	344	82	30	18	7154	211	53
17	53	56	34	60	121	379	590	169	26	240	6159	361
19	21	25	43	51	54	84	629	482	32	42	344	4852

The entry $\mathbf{C}_S(j, k)$ of this matrix denotes the number of feature correspondences found when all descriptors derived from image $\mathscr{I}_{1,j}$ are compared to the respective nearest and second nearest neighbors derived from image $\mathscr{I}_{1,k}$

for a SIFT descriptor derived for a keypoint in one image $\mathscr{I}_{\mathrm{I},j}$, the nearest SIFT descriptor and the second nearest SIFT descriptor in the other image $\mathscr{I}_{\mathrm{I},k}$ are a little more distinctive as required, this ratio ρ is below the threshold t_{des} and thus meets the constraint. In the reverse case, when comparing a feature descriptor derived from image $\mathscr{I}_{\mathrm{I},k}$ to feature descriptors derived from image $\mathscr{I}_{\mathrm{I},j}$, it might occur that the nearest SIFT descriptor and the second nearest SIFT descriptor in image $\mathscr{I}_{\mathrm{I},j}$ are more similar which results in a ratio $\rho > t_{\mathrm{des}}$. As the confusion matrix $\mathbf{C}_S$ is not necessarily symmetric, we use a directed graph instead of an undirected graph. Hence, the entry $\mathbf{C}_S(j,k)$ of the confusion matrix represents the weight of an unidirectional edge from node $\mathscr{S}_j$ to node $\mathscr{S}_k$.

The first step toward organizing the TLS scans consists of an initialization which may be done by selecting a reference scan (which is represented by scan $\mathscr{S}_{01}$ in our case). Alternatively, it would be possible to use other criteria if only the relations between the scans are of importance, e.g., the node from which the edge with the maximum weight within the graph starts. The reference scan thus represents an initial set containing exactly one node, and this initial set is iteratively expanded until it contains all nodes of the graph. Each iteration starts with a search for unidirectional edges from the current set of nodes to the remaining nodes and the edge with the maximum weight leads to the node by which the current set is expanded. Resulting from the selected connections, a structure may be generated which represents the order of the scans for a successive automatic pair-wise registration scheme. For the exemplary confusion matrix provided in Table 4.3 where the scan IDs represent the original scan IDs provided with the given TLS dataset, the resulting structure for a successive pair-wise registration process is shown in Fig. 4.9 and the derived scan order for a successive pair-wise registration indeed coincides with the scan IDs proposed in [21] as shown in Fig. 4.10. The whole procedure for organizing the scans based on the use of SIFT features extracted via the original SIFT implementation [53, 54] takes approximately 607.04 s for the given set of 12 scans on a standard desktop computer (Intel Core2 Quad Q9550, 2.83 GHz, 8 GB RAM, Matlab implementation). The derived scan order is the same when for instance using SURF features as the basis for defining the edge weights in the graph.

4.4.3.2 Projective Scan Matching

Once the scans have been ordered for a successive pair-wise registration scheme, we may have a closer look on the registration results in terms of accuracy and efficiency. For this purpose, we first focus on using SIFT features in order to detect 2D/2D feature correspondences ($t_{\mathrm{des}} = 0.66$), our projective scan matching technique relying on the RANSAC-based EPnP scheme for coarse registration, a geometric outlier removal based on the measure $\tilde{d}$, and a subsequent refinement via a repetition of the RANSAC-based EPnP scheme. For this purpose, we consider the scan pairs $\mathscr{P}_j$ with $j = 1, \ldots, 10$ which provide a spacing of about 4–6 m between the respective scan positions (Fig. 4.12 and Table 4.4) and a number of 217–656 feature correspondences detected between the respective intensity images (Fig. 4.11 and Table 4.5). The

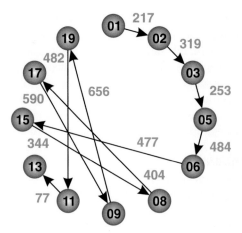

Fig. 4.9 Resulting graph for a successive pair-wise registration: the nodes representing the respective scans are labeled with the original scan ID and the connections used for a successive pair-wise registration are labeled with the number of detected feature correspondences between the respective intensity images

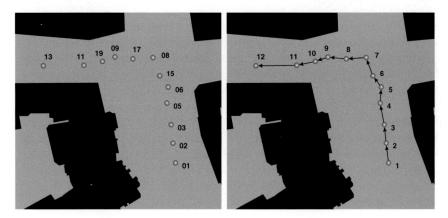

Fig. 4.10 Map of the Hannover "Holzmarkt" with the original scan IDs provided with the dataset (*left*) and the modified scan IDs assigned to the different scans $\mathscr{S}_j$ (*right*). Note that the modified scan IDs are in accordance with [21] as shown in Fig. 4.8

resulting position and angle errors after coarse registration are provided in Table 4.6, and the respective errors after the refinement are provided in Table 4.7. These tables reveal that the coarse registration yields an absolute position error in the range between 16 and 42 mm (Table 4.6, last column), and the refinement leads to an absolute position error in the range between 8 and 29 mm (Table 4.7, last column) which is even below the given measurement accuracy of 12 mm. The respective angle errors after the refinement are below 0.065° in almost all cases; only one value is larger with 0.079° (Table 4.7). Note that the refinement results in an improvement

Fig. 4.11 Visualization of the detected SIFT features when considering the scans $\mathscr{S}_1$ and $\mathscr{S}_2$ (*top*) and the derived 217 feature correspondences when using a threshold of $t_{\mathrm{des}} = 0.66$ for feature matching (*bottom*)

with respect to the estimated scan position as shown in Fig. 4.13. However, it has to be stated that in cases where the coarse estimate already shows a high accuracy, the refinement may lead to an improvement, but not necessarily. In some cases, despite an improvement of the position estimate, single rotation angles are affected in a negative way (Tables 4.6 and 4.7). In order to obtain an impression about the computational effort for pair-wise registration, we may report an average processing time of about 5 s for the feature matching and a total time of 13 s for coarse registration, outlier removal, and refinement on a standard desktop computer (Intel Core2 Quad Q9550, 2.83 GHz, 8 GB RAM, Matlab implementation).

Table 4.4 Absolute 3D distance (in m) between the single scan positions indicated with the assigned scan IDs

ID	1	2	3	4	5	6	7	8	9	10	11
1	0	5.58	10.85	16.95	21.47	24.86	30.40	31.79	34.45	35.55	38.32
2	5.58	0	5.27	11.37	15.89	19.30	24.85	26.46	29.41	30.82	34.02
3	10.85	5.27	0	6.11	10.62	14.05	19.62	21.54	24.86	26.66	30.34
4	16.95	11.37	6.11	0	4.60	7.95	13.53	15.93	19.78	22.12	26.41
5	21.47	15.89	10.62	4.60	0	3.96	9.42	12.93	17.48	20.42	25.24
6	24.86	19.30	14.05	7.95	3.96	0	5.59	9.11	13.90	17.11	22.18
7	30.40	24.85	19.62	13.53	9.42	5.59	0	5.81	10.97	14.77	20.16
8	31.79	26.46	21.54	15.93	12.93	9.11	5.81	0	5.18	8.95	14.35
9	34.45	29.41	24.86	19.78	17.48	13.90	10.97	5.18	0	3.92	9.31
10	35.55	30.82	26.66	22.12	20.42	17.11	14.77	8.95	3.92	0	5.40
11	38.32	34.02	30.34	26.41	25.24	22.18	20.16	14.35	9.31	5.40	0

Fig. 4.12 View on the 3D point clouds after their alignment in a common coordinate frame via the reference data

4.4.3.3   Influence of Point Quality Assessment on Projective Scan Matching

In this subsection, we present the results obtained when exploiting SIFT features in order to detect 2D/2D feature correspondences ($t_{des} = 0.66$), our projective scan matching technique relying on the RANSAC-based EPnP scheme for coarse

Table 4.5 Number of feature correspondences between the scans indicated with the assigned scan IDs

ID	1	2	3	4	5	6	7	8	9	10	11
1	4986	217	63	45	33	44	58	62	41	39	28
2	229	5663	319	100	59	35	80	46	43	38	48
3	88	308	5967	253	120	57	56	38	47	56	68
4	70	114	277	6200	484	131	78	68	58	84	68
5	31	70	124	466	6682	477	169	134	68	71	56
6	61	40	59	129	503	7154	344	211	82	53	30
7	86	96	53	78	163	328	6867	404	205	99	64
8	53	56	34	60	121	240	379	6159	590	361	169
9	39	34	37	56	68	78	158	577	5571	656	330
10	21	25	43	51	54	42	84	344	629	4852	482
11	17	24	37	40	44	24	41	134	277	408	4061

Table 4.6 Errors between the estimated transformation parameters after coarse registration and the reference data

Scan pair	$\Delta\omega$ (°)	$\Delta\phi$ (°)	$\Delta\kappa$ (°)	ΔX (m)	ΔY (m)	ΔZ (m)	e_{pos} (m)
$\mathscr{P}_1$	−0.001	−0.041	0.057	−0.023	0.030	−0.002	0.038
$\mathscr{P}_2$	0.009	0.007	0.053	−0.011	−0.014	−0.003	0.018
$\mathscr{P}_3$	−0.014	−0.072	0.029	0.015	0.024	−0.022	0.036
$\mathscr{P}_4$	0.002	−0.007	0.002	0.016	−0.025	−0.003	0.030
$\mathscr{P}_5$	0.038	−0.009	0.003	−0.004	−0.008	−0.013	0.016
$\mathscr{P}_6$	0.046	0.035	0.018	0.012	−0.023	0.005	0.026
$\mathscr{P}_7$	0.037	−0.000	−0.017	0.027	−0.013	0.010	0.032
$\mathscr{P}_8$	−0.013	−0.002	0.031	−0.017	−0.010	0.007	0.021
$\mathscr{P}_9$	−0.067	−0.009	0.043	−0.035	0.008	−0.022	0.042
$\mathscr{P}_{10}$	0.025	−0.064	−0.011	0.037	−0.001	0.007	0.038

Table 4.7 Errors between the estimated transformation parameters after fine registration and the reference data

Scan pair	$\Delta\omega$ (°)	$\Delta\phi$ (°)	$\Delta\kappa$ (°)	ΔX (m)	ΔY (m)	ΔZ (m)	e_{pos} (m)
$\mathscr{P}_1$	−0.047	−0.011	0.004	0.019	−0.001	−0.005	0.019
$\mathscr{P}_2$	0.044	−0.013	0.035	−0.004	−0.009	−0.003	0.010
$\mathscr{P}_3$	0.034	−0.004	0.079	−0.015	−0.006	0.010	0.019
$\mathscr{P}_4$	−0.021	0.022	0.008	−0.002	−0.012	0.005	0.013
$\mathscr{P}_5$	0.031	−0.002	0.009	−0.001	−0.005	−0.008	0.010
$\mathscr{P}_6$	0.017	0.032	0.053	0.004	−0.011	0.010	0.015
$\mathscr{P}_7$	0.054	−0.044	0.012	0.002	0.006	−0.005	0.008
$\mathscr{P}_8$	0.015	−0.027	0.031	0.009	0.005	−0.008	0.013
$\mathscr{P}_9$	−0.043	−0.004	0.029	−0.008	0.003	−0.012	0.015
$\mathscr{P}_{10}$	0.016	−0.065	−0.014	0.025	−0.012	0.009	0.029

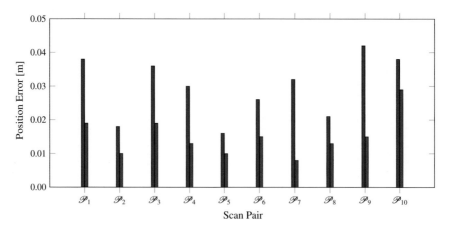

Fig. 4.13 Position error after coarse registration (*blue*) and after the refinement (*red*)

registration, a geometric outlier removal based on the measure $\tilde{d}$, and a subsequent ICP-based fine registration. Thereby, we also involve the different methods for point quality assessment in order to remove unreliable feature correspondences ($t_{I,gray} = 10$, $t_\sigma = 0.10$ m). Since the random sampling may lead to slightly different estimates, we average all position and angle estimates over 20 runs. For the different scan pairs $\mathscr{P}_j = \{\mathscr{S}_j, \mathscr{S}_{j+1}\}$ with $j = 1, \ldots, 11$, the remaining position errors after coarse and fine registration are shown in Fig. 4.14 as well as the achieved improvement. These results reveal that already the step of coarse registration provides accurate position estimates, where the position error indicating the absolute deviation of the estimated scan position from the reference data is less than 5 cm for almost all cases. After fine registration, the remaining position error is in the range between 0.47 cm and 4.10 cm. The respective angle errors indicating the deviation of the estimated angles from the reference angles are in the interval between $0.0001°$ and $0.2845°$ after coarse registration, and they are reduced to the interval between $0.0002°$ and $0.0919°$ after fine registration.

In order to obtain an impression about the computational effort for a pair-wise registration, the mean processing times required for the different subtasks on a standard desktop computer (Intel Core2 Quad Q9550, 2.83 GHz, 8 GB RAM, Matlab implementation) are listed in Table 4.8. Since those processing times for coarse registration vary significantly when involving different methods for point quality assessment, a respective visualization is provided in Fig. 4.15. Based on these numbers, in total, a processing time of 191.35 s may be expected in the worst case for a pair-wise registration of the considered TLS scans.

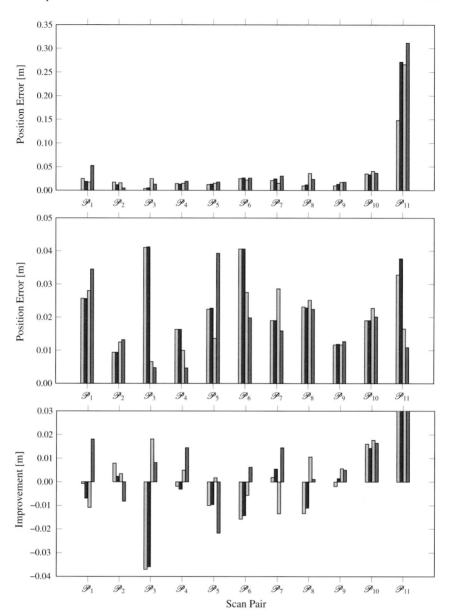

Fig. 4.14 Mean position error after coarse registration (*top*), after fine registration (*center*), and the respective improvement (*bottom*) for the scan pairs $\mathscr{P}_j = \left\{\mathscr{S}_j, \mathscr{S}_{j+1}\right\}$ when applying no reliability check (*gray*) and when applying reliability checks with respect to intensity information (*magenta*), range reliability (*green*), or local planarity (*teal*)

Table 4.8 Required processing times for the different subtasks of a pair-wise registration on a standard desktop computer (Intel Core2 Quad Q9550, 2.83 GHz, 8 GB RAM, Matlab implementation)

Task	Number of executions	Processing time (s)
Point quality assessment	2	
–Intensity information		<0.01
–Range reliability		1.32
–Local planarity		78.39
Feature extraction	2	8.77
Forward projection	2	0.17
Feature matching	1	0.21
Coarse registration	1	0.10 . . . 16.37
Outlier removal	1	<0.04
Fine registration	1	<0.07

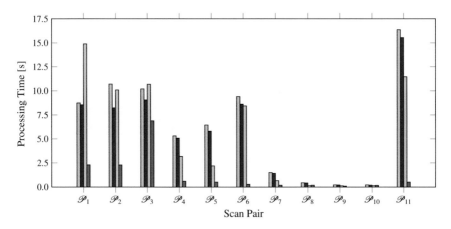

Fig. 4.15 Mean processing times required for the coarse registration of scan pairs $\mathcal{P}_j = \{\mathcal{S}_j, \mathcal{S}_{j+1}\}$ when applying no reliability check (*gray*) and reliability checks with respect to intensity information (*magenta*), range reliability (*green*), or local planarity (*teal*)

4.4.3.4 Omnidirectional Scan Matching Based on Different Feature Detector–Descriptor Combinations

In this subsection, we present the results obtained when solving the relative orientation problem by aligning sets of bearing vectors as described in Sect. 4.3.6.2. Again, we focus on a successive pair-wise registration of scan pairs $\mathcal{P}_j = \{\mathcal{S}_j, \mathcal{S}_{j+1}\}$ with $j = 1, \ldots, 11$, but we additionally intend to quantify the impact of the involved features and therefore consider the different 2D keypoint detector–descriptor combinations as described in Sect. 4.3.3. In order to obtain comparable results, we use the implementations provided in OpenCV 2.4 for SIFT, SURF, and ORB, while we use the implementations released with the respective paper for the other two

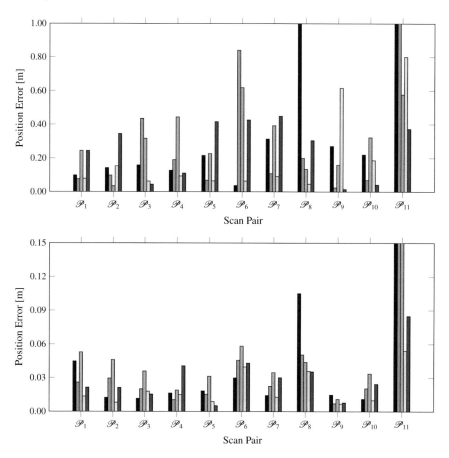

Fig. 4.16 Position errors after the coarse registration (*top*) and after the fine registration (*bottom*) of scan pairs $\mathscr{P}_j = \{\mathscr{S}_j, \mathscr{S}_{j+1}\}$: SIFT (*violet*), SURF (*cyan*), ORB (*green*), A-KAZE + M-SURF (*yellow*), SURF* + BinBoost (*red*)

keypoint detector–descriptor combinations represented by A-KAZE + M-SURF and SURF* + BinBoost.

The respective position and angle errors after coarse and fine registration are visualized in Figs. 4.16 and 4.17, whereby the position error indicates the deviation of the estimated scan position from the reference values and the angle error has been determined as the difference of respective Rodrigues vectors with respect to their length [47]. Furthermore, we provide the number of 2D/2D feature correspondences used for coarse and fine registration in Fig. 4.18 in order to quantify differences between the different methods for feature extraction and matching. For coarse registration, we additionally provide the ratio of inliers with respect to all feature correspondences as well as the number of RANSAC iterations in Fig. 4.19. In order to obtain an impression about the computational effort on a standard notebook

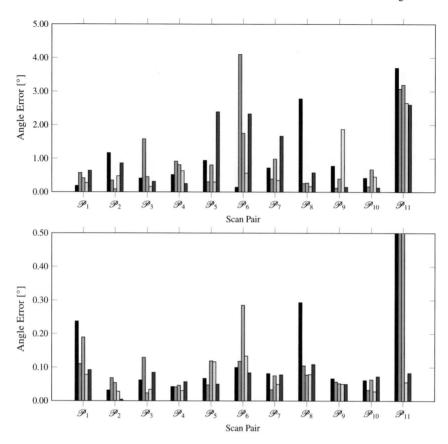

Fig. 4.17 Average angle errors after the coarse registration (*top*) and after the fine registration (*bottom*) of scan pairs $\mathscr{P}_j = \{\mathscr{S}_j, \mathscr{S}_{j+1}\}$: SIFT (*violet*), SURF (*cyan*), ORB (*green*), A-KAZE + M-SURF (*yellow*), SURF* + BinBoost (*red*)

(Intel Core i7-3630QM, 2.4 Ghz, 16 GB RAM), the average processing times for the different subtasks are provided in Table 4.9 as well as the expected time for the whole process of aligning two scans. Finally, we also provide a visualization of registered TLS scans in Fig. 4.20.

4.5 Discussion

Generally, we may state that our framework is well-suited for both scenes representing urban environments and scenes containing vegetation, and it does neither depend on the presence of regular surfaces nor require human interaction. In comparison to approaches relying on features directly extracted from the considered 3D

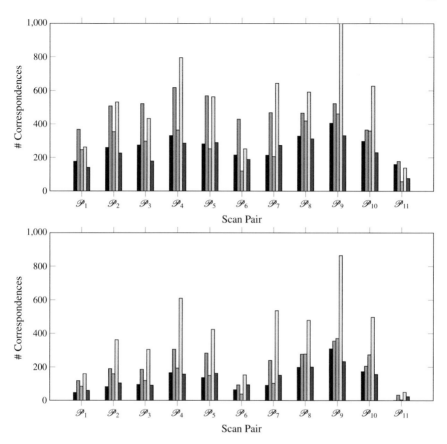

Fig. 4.18 Number of feature correspondences used for the coarse registration (*top*) and for the fine registration (*bottom*) of scan pairs $\mathscr{P}_j = \{\mathscr{S}_j, \mathscr{S}_{j+1}\}$: SIFT (*violet*), SURF (*cyan*), ORB (*green*), A-KAZE + M-SURF (*yellow*), SURF* + BinBoost (*red*)

point clouds, the use of local features extracted from the 2D image representations derived for the respective scans allows a robust, efficient, and accurate alignment of scans in a common coordinate frame. Thus, the strategy of 2D keypoint-based point cloud registration cuts down the computational burden for the registration of larger 3D point clouds consisting of millions of 3D points to a few seconds on a standard computer, while preserving a fully automated and markerless registration without involving any prior knowledge about the scene.

The only limitation of our framework may be identified in the fact that 2D/2D feature correspondences have to be derived between the intensity images derived for the respective scans. In this regard, the proposed framework is suited for scenes with well-textured object surfaces, whereas it will probably not lead to optimal results for scenes with less textured object surfaces. Furthermore, we may generally observe that the total number of feature correspondences decreases with an increasing distance

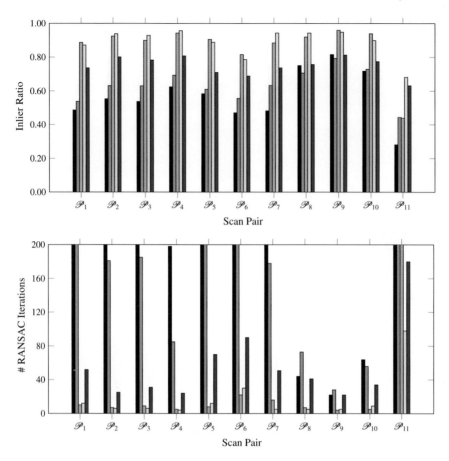

Fig. 4.19 Inlier ratio (*top*) and number of RANSAC iterations (*bottom*) for the coarse registration of scan pairs $\mathscr{P}_j = \{\mathscr{S}_j, \mathscr{S}_{j+1}\}$: SIFT (*violet*), SURF (*cyan*), ORB (*green*), A-KAZE + M-SURF (*yellow*), SURF* + BinBoost (*red*)

between the respective scan positions (Tables 4.5 and 4.4). Consequently, the quality of the registration results will decrease for larger distances between considered scans. However, this constraint holds for the other image-based approaches as well and is not specific for our framework. For larger distances between the considered scans, approaches relying on geometric primitives as meaningful features might be more robust which, for instance, becomes visible when considering results derived for registration approaches relying on the use of planes as features [21, 96]. Yet, such approaches are based on the assumption that specific regular surfaces are given in the observed scene and they are therefore less general concerning the scene structure which, in turn, represents a significant disadvantage.

Table 4.9 Average processing times t_{FEX} for feature extraction, t_{FM} for feature matching, t_{CR} for coarse registration, t_{FR} for fine registration, and average total time t_Σ required for automatically aligning two scans

Method	t_{FEX} (s)	t_{FM} (s)	t_{CR} (s)	t_{FR} (s)	t_Σ (s)
SIFT	3.055	10.479	0.084	0.012	16.684
SURF	0.644	0.414	0.081	0.022	1.805
ORB	0.138	2.023	0.015	0.021	2.336
A-KAZE + M-SURF	2.815	2.533	0.007	0.050	8.221
SURF* + BinBoost	2.423	0.067	0.025	0.013	4.950

Fig. 4.20 Aligned 3D point clouds when using A-KAZE + M-SURF: the 3D points belonging to different scans $\mathscr{S}_j$ are encoded in different colors

As most competing strategies to 2D keypoint-based point cloud registration, we may consider the strategy of 3D keypoint-based point cloud registration [86–88] which—due to the use of a line-of-sight instrument with a specific angular resolution—typically involves a voxelization of the scene in order to cope with the significant variation in point density and respectively extract 3D keypoints. Such a strategy transforming the measured 3D point cloud data to a subsampled representation of the data, however, strongly depends on the selected voxel size. A further competitive strategy relies on obtaining viewpoint invariance by means of an orthographic projection along detected salient directions in the range data (e.g., in the form of peaks in the distribution of surface normals, vanishing points, symmetry, gravity, or other characteristic directions) and has recently been presented [109]. This strategy focuses on extracting local features from salient direction rectified images (of a virtual orthographic camera), and it is reported to also allow a registration of scans

with a larger distance between the respective scan positions and thus less overlap
between the respective scans.

Projective Scan Matching When transferring the task of point cloud registration to
the task of solving the PnP problem, we may generally observe an already relatively
good position estimate after coarse registration except for the last scan pair $\mathscr{P}_{11}$
(Fig. 4.14, top). This however results from the fact that the distance between the
respective scans is between 3.93 and 6.11 m for the scan pairs $\mathscr{P}_1, \ldots, \mathscr{P}_{10}$, whereas
it is almost 12 m for scan pair $\mathscr{P}_{11}$ (Table 4.2). Due to the significantly larger distance,
the similarity between the respective intensity images becomes less (Table 4.5) and,
consequently, the number of derived 2D/2D feature correspondences decays quickly
compared to the other scan pairs. After fine registration, however, the remaining
position error for scan pair $\mathscr{P}_{11}$ is reduced to the same range as for the other scan
pairs (Fig. 4.14, center), and this behavior also holds for the respective angle errors.

When having a closer look on the results for the involved methods for point quality assessment, we may state that a filtering of feature correspondences via intensity
information has only very little influence on the registration results (Fig. 4.14, center),
while the proposed measure $\sigma_{r,3\times3}$ of range reliability mainly improves the results for
those scan pairs where the position error tends to be larger (e.g., $\mathscr{P}_3$, $\mathscr{P}_6$ and $\mathscr{P}_{11}$).
Even a relatively strict filtering of feature correspondences based on our proposed
measure $P_{\lambda,3\times3}$ of local planarity does not always lead to an improvement after fine
registration (Fig. 4.14, center). However, this may be due to the fact that the accuracy
after fine registration is quite close to the expected measurement accuracy of the scanning device (12 mm). In this regard, it may be taken into account that the RANSAC-
based EPnP scheme involves both 3D and 2D cues, and thus already ensures a relatively reliable coarse registration compared to approaches only focusing on spatial
3D geometry, where the novel point quality measures may show a more significant
improvement of the registration results. Interestingly, however, the main effect of
using the measure $P_{\lambda,3\times3}$ of local planarity for point quality assessment consists
of a significant speedup in coarse registration (Fig. 4.15), while causing additional
costs in point quality assessment compared to the other methods (Table 4.8). The
speedup in coarse registration, in turn, is important since a fast solution corresponds
to a reliable estimate of the relative orientation between two scans. More specifically, a filtering of feature correspondences based on the proposed measure $P_{\lambda,3\times3}$
of local planarity represents a consistency check that—like specific modifications of
RANSAC [76]—results in a reduced set of feature correspondences, where the inlier
ratio is significantly increased which, in turn, leads to a faster convergence of the
RANSAC algorithm toward a suitable solution. Thereby, the generic consideration
of incidence angles up to about 45° (Sect. 2.5.6) imposes more restrictions than other
recent investigations addressing an optimized selection of scan positions [78], where
incidence angles up to 70° are assumed to result in reliable range measurements.

Omnidirectional Scan Matching When transferring the task of point cloud registration to the task of finding the transformation between sets of bearing vectors, the
results provided in Figs. 4.16 and 4.17 reveal that the position errors after fine registration are less than 0.06 m for almost all keypoint detector–descriptor combinations

when considering the scan pairs $\mathscr{P}_1, \ldots, \mathscr{P}_{10}$, where the distance between the respective scan positions is between 3.93 and 6.11 m. Since the respective angle errors (i.e., the difference of respective Rodrigues vectors) after fine registration are below 0.15° with only a few exceptions, we may conclude that the presented method for coarse registration represents a competitive method in order to coarsely align the given scans and that a respective outlier rejection based on Euclidean 3D distances is sufficient for an ICP-based fine registration. The applicability of our method for coarse registration is even further motivated by the fact that the respective processing time is less than 0.085 s for the considered scan pairs (Table 4.9). Thus, we reach a total time of less than 10 s for the registration of the considered scan pairs for four of the five tested keypoint detector–descriptor combinations, and only the involved implementation for SIFT is not that efficient. Note that the time required for feature extraction is counted twice, since this task is required for both scans of a scan pair.

Considering the five involved keypoint detector–descriptor combinations, we may state that A-KAZE + M-SURF and SURF* + BinBoost tend to provide the best results after fine registration (Fig. 4.16 and Fig. 4.17). Note that only these combinations are also able to derive a suitable position and angle estimate for the last scan pair $\mathscr{P}_{11} = \{\mathscr{S}_{11}, \mathscr{S}_{12}\}$, where the distance between the respective scan positions is approximately 12 m. The respective position errors for A-KAZE + M-SURF and SURF* + BinBoost after fine registration are 0.054 and 0.085 m, while the angle errors are 0.056° and 0.083°, respectively. In contrast, SIFT, SURF, and ORB provide a position error of more than 0.20 m and an angle error of more than 0.75° for that case.

In Fig. 4.18, it becomes visible that the number of feature correspondences used for coarse registration is similar for SIFT, ORB, and SURF* + BinBoost, while it tends to be higher for SURF. For A-KAZE + M-SURF, even a significant increase of this number may be observed across all 12 scan pairs. The increase in the number of involved feature correspondences for A-KAZE + M-SURF compared to the other keypoint detector–descriptor combinations is even more significant when considering fine registration, where it is partially even more than twice as much as for the others. Based on these characteristics, an interesting trend becomes visible when considering the respective ratio of inliers during coarse registration. While the inlier ratio is comparable for SIFT and SURF, it is better for SURF* + BinBoost, and it is considerably better for ORB and A-KAZE + M-SURF (Fig. 4.19, top). A high percentage of inliers, in turn, has a positive impact on coarse registration by significantly reducing the number of RANSAC iterations (Fig. 4.19, bottom). Consequently, the combination A-KAZE + M-SURF not only increases the number of feature correspondences, but also the inlier ratio and, thus, the position and angle errors across all scan pairs tend to be the lowest for this combination (Figs. 4.16 and 4.17). For most of the scan pairs, the respective position error is even close or below the given measurement accuracy of 12 mm.

4.6 Conclusions

In this chapter, we have addressed the adequate alignment of several, unordered 3D point clouds in a common coordinate frame which represents an important prerequisite for the digitization and 3D reconstruction of real-world scenes. Particular interest has been paid to fully automated, markerless, efficient, and robust approaches for a pair-wise registration of 3D point clouds. In this regard, a key component has been identified in local features extracted from 2D imagery. These local features represented in the form of 2D keypoints and respective keypoint descriptors allow an efficient detection of corresponding contents in the intensity images derived for the respective 3D point clouds. Together with the respective range information, these 2D keypoints may be exploited in order to derive sparse 3D point sets of physically (almost) identical points in 3D space. While keypoint-based point cloud registration typically focuses on directly aligning the sparse 3D point sets (e.g., via estimating a rigid transformation in the least squares sense) and hence only considers 3D cues, we have instead focused on two different strategies for keypoint-based point cloud registration. The first strategy, referred to as projective scan matching, focuses on the consideration of both 3D cues in terms of how good corresponding 3D points fit together and 2D cues in terms of how good the 2D locations in the respective image representations fit together. As a result, we have obtained highly accurate registration results already for coarse registration. The second strategy, referred to as omnidirectional scan matching, is based on the idea that 2D keypoints may be detected with subpixel accuracy, while the range measurements tend to be corrupted with more or less noise, depending on different influencing factors. Consequently, we have exploited corresponding 2D keypoints and the respective range information in order to derive sets of bearing vectors which have been accurately aligned by using the well-known five-point algorithm.

While the applied strategy for keypoint-based point cloud registration certainly has the strongest impact on the quality of the derived registration results, we furthermore focused on checking the quality of scanned 3D points in order to remove unreliable feature correspondences. In this regard, we have tested two novel measures represented by range reliability and local planarity. While the measure of range reliability only takes into account the geometric smoothness of object surfaces, the measure of local planarity also takes into account the respective incidence angle of the local object surface. Particularly when applying a filtering based on the measure of local planarity, the ratio of inlier correspondences is increased which, in turn, leads to a faster convergence of the RANSAC algorithm toward a suitable solution. Additionally, we have evaluated the influence of different 2D keypoint detector–descriptor combinations on the results of point cloud registration. In this regard, the derived results clearly reveal that replacing SIFT and SURF detectors and descriptors by more recent approaches significantly alleviates point cloud registration in terms of robustness, efficiency, and accuracy.

For future work, it would be desirable to compare different approaches for point cloud registration on several benchmark datasets and to point out chances and limitations of these approaches with respect to different criteria specified by potential end-users, e.g., the spacing between adjacent scans or the complexity of the observed scene, in order to allow end-users to select an appropriate method according to their requirements. Thereby, further approaches for feature extraction should be taken into consideration as well as schemes for weighting the influence of the derived feature correspondences on the registration results. For the latter, different criteria based on point quality measures, the similarity of corresponding features or further constraints could be evaluated. Furthermore, a global registration over all scans or at least those scans with a high similarity according to the derived confusion matrix should be taken into consideration, since this might improve the quality of the estimated transformation parameters and yield an even further increased quality of the registration results.

References

1. Agrawal M, Konolige K, Blas MR (2008) CenSurE: center surround extremas for realtime feature detection and matching. In: Proceedings of the European conference on computer vision, vol IV, pp 102–115
2. Aiger D, Mitra NJ, Cohen-Or D (2008) 4-points congruent sets for robust pairwise surface registration. ACM Trans Graph 27(3):1–10
3. Akca D (2003) Full automatic registration of laser scanner point clouds. In: Proceedings of optical 3-d measurement techniques VI, vol I, pp 330–337
4. Al-Durgham M, Detchev I, Habib A (2011) Analysis of two triangle-based multi-surface registration algorithms of irregular point clouds. Int Arch Photogramm Remote Sens Spat Inf Sci XXXVIII-5/W12:61–66
5. Al-Manasir K, Fraser CS (2006) Registration of terrestrial laser scanner data using imagery. Photogramm Rec 21(115):255–268
6. Alba M, Barazzetti L, Scaioni M, Remondino F (2011) Automatic registration of multiple laser scans using panoramic RGB and intensity images. Int Arch Photogramm Remote Sens Spat Inf Sci XXXIII-5/W12:49–54
7. Alcantarilla PF, Bartoli A, Davison AJ (2012) KAZE features. In: Proceedings of the European conference on computer vision, vol VI, pp 214–227
8. Alcantarilla PF, Nuevo J, Bartoli A (2013) Fast explicit diffusion for accelerated features in nonlinear scale spaces. In: Proceedings of the British machine vision conference, pp 13.1–13.11
9. Arun KS, Huang TS, Blostein SD (1987) Least-squares fitting of two 3-D point sets. IEEE Trans Pattern Anal Mach Intell 9(5):698–700
10. Attneave F (1954) Some informational aspects of visual perception. Psychol Rev 61(3):183–193
11. Bae K-H, Belton D, Lichti DD (2005) A framework for position uncertainty of unorganised three-dimensional point clouds from near-monostatic laser scanners using covariance analysis. Int Arch Photogramm Remote Sens Spat Inf Sci XXXVI-3/W19:7–12
12. Bae K-H, Lichti DD (2008) A method for automated registration of unorganised point clouds. ISPRS J Photogramm Remote Sens 63(1):36–54
13. Barnea S, Filin S (2007) Registration of terrestrial laser scans via image based features. Int Arch Photogramm Remote Sens Spat Inf Sci XXXVI-3/W52:32–37
14. Barnea S, Filin S (2008) Keypoint based autonomous registration of terrestrial laser point-clouds. ISPRS J Photogramm Remote Sens 63(1):19–35

15. Bay H, Tuytelaars T, Van Gool L (2006) SURF: speeded up robust features. In: Proceedings of the European conference on computer vision, vol 1, pp 404–417
16. Bay H, Ess A, Tuytelaars T, Van Gool L (2008) Speeded-up robust features (SURF). Comput Vis Image Underst 110(3):346–359
17. Bendels GH, Degener P, Wahl R, Körtgen M, Klein R (2004) Image-based registration of 3D-range data using feature surface elements. In: Proceedings of the international symposium on virtual reality, archaeology and cultural heritage, pp 115–124
18. Besl PJ, McKay ND (1992) A method for registration of 3-D shapes. IEEE Trans Pattern Anal Mach Intell 14(2):239–256
19. Biber P, Strasser W (2003) The normal distributions transform: a new approach to laser scan matching. In: Proceedings of the IEEE/RSJ international conference on intelligent robots and systems, vol 3, pp 2743–2748
20. Boehm J, Becker S (2007) Automatic marker-free registration of terrestrial laser scans using reflectance features. In: Proceedings of optical 3-d measurement techniques VIII, pp 338–344
21. Brenner C, Dold C, Ripperda N (2008) Coarse orientation of terrestrial laser scans in urban environments. ISPRS J Photogramm Remote Sens 63(1):4–18
22. Bresenham JE (1965) Algorithm for computer control of a digital plotter. IBM Syst J 4(1):25–30
23. Calonder M, Lepetit V, Strecha C, Fua P (2010) BRIEF: binary robust independent elementary features. In: Proceedings of the European conference on computer vision, vol IV, pp 778–792
24. Censi A (2008) An ICP variant using a point-to-line metric. In: Proceedings of the IEEE international conference on robotics and automation, pp 19–25
25. Chen Y, Medioni G (1992) Object modelling by registration of multiple range images. Image Vis Comput 10(3):145–155
26. Dahl AL, Aanæs H, Pedersen KS (2011) Finding the best feature detector-descriptor combination. In: Proceedings of the IEEE international conference on robotics and automation, pp 318–325
27. Dold C (2005) Extended Gaussian images for the registration of terrestrial scan data. Int Arch Photogramm Remote Sens Spat Inf Sci XXXVI-3/W19:180–185
28. Eggert DW, Lorusso A, Fisher RB (1997) Estimating 3-D rigid body transformations: a comparison of four major algorithms. Mach Vis Appl 9(5–6):272–290
29. Fischler MA, Bolles RC (1981) Random sample consensus: a paradigm for model fitting with applications to image analysis and automated cartography. Commun ACM 24(6):381–395
30. Forkuo EK, King B (2004) Automatic fusion of photogrammetric imagery and laser scanner point clouds. Int Arch Photogramm Remote Sens Spat Inf Sci XXXV-B4:921–926
31. Franaszek M, Cheok GS, Witzgall C (2009) Fast automatic registration of range images from 3D imaging systems using sphere targets. Autom Constr 18(3):265–274
32. Gressin A, Mallet C, Demantké J, David N (2013) Towards 3D lidar point cloud registration improvement using optimal neighborhood knowledge. ISPRS J Photogramm Remote Sens 79:240–251
33. Gruen A, Akca D (2005) Least squares 3D surface and curve matching. ISPRS J Photogramm Remote Sens 59(3):151–174
34. Hänsch R, Weber T, Hellwich O (2014) Comparison of 3D interest point detectors and descriptors for point cloud fusion. ISPRS Ann Photogramm Remote Sens Spat Inf Sci II-3:57–64
35. Hamming RW (1950) Error detecting and error correcting codes. Bell Syst Tech J 29(2):147–160
36. Harris C, Stephens M (1988) A combined corner and edge detector. In: Proceedings of the Alvey vision conference, pp 147–151
37. Hartley RI, Zisserman A (2008) Multiple view geometry in computer vision. University Press, Cambridge
38. Hebel M, Stilla U (2007) Automatic registration of laser point clouds of urban areas. Int Arch Photogramm Remote Sens Spat Inf Sci XXXVI-3/W49A:13–18
39. Hebel M, Stilla U (2009) Automatische Koregistrierung von ALS-Daten aus mehreren Schräg-ansichten städtischer Quartiere. PFG—Photogramm Fernerkund Geoinf 3(2009):261–275

40. Hebel M, Stilla U (2012) Simultaneous calibration of ALS systems and alignment of multiview lidar scans of urban areas. IEEE Trans Geosci Remote Sens 50(6):2364–2379
41. Horn BKP (1984) Extended Gaussian images. Proc IEEE 72(12):1671–1686
42. Horn BKP, Hilden HM, Negahdaripour S (1988) Closed-form solution of absolute orientation using orthonormal matrices. J Opt Soc Am A 5(7):1127–1135
43. Huang J, You S (2012) Point cloud matching based on 3D self-similarity. In: Proceedings of the IEEE computer society conference on computer vision and pattern recognition workshops, pp 41–48
44. Huber DF, Hebert M (2003) Fully automatic registration of multiple 3D data sets. Image Vis Comput 21(7):637–650
45. Kang Z, Li J, Zhang L, Zhao Q, Zlatanova S (2009) Automatic registration of terrestrial laser scanning point clouds using panoramic reflectance images. Sensors 9(4):2621–2646
46. Khoshelham K, Dos Santos DR, Vosselman G (2013) Generation and weighting of 3D point correspondences for improved registration of RGB-D data. ISPRS Ann Photogramm Remote Sens Spat Inf Sci II-5/W2:127–132
47. Kneip L, Furgale P (2014) OpenGV: a unified and generalized approach to real-time calibrated geometric vision. In: Proceedings of the IEEE international conference on robotics and automation, pp 1–8
48. Lepetit V, Moreno-Noguer F, Fua P (2009) EPnP: an accurate $O(n)$ solution to the PnP problem. Int J Comput Vis 81(2):155–166
49. Li J, Allinson NM (2008) A comprehensive review of current local features for computer vision. Neurocomputing 71(10–12):1771–1787
50. Lindeberg T (1994) Scale-space theory: a basic tool for analysing structures at different scales. J Appl Stat 21(2):224–270
51. Lindeberg T (1998) Feature detection with automatic scale selection. Int J Comput Vis 30(2):79–116
52. Longuet-Higgins HC (1987) A computer algorithm for reconstructing a scene from two projections. In: Fischler MA, Firschein O (eds) Readings in computer vision: issues, problems, principles, and paradigms. Morgan Kaufmann, San Francisco, pp 61–62
53. Lowe DG (1999) Object recognition from local scale-invariant features. In: Proceedings of the IEEE international conference on computer vision, pp 1150–1157
54. Lowe DG (2004) Distinctive image features from scale-invariant keypoints. Int J Comput Vis 60(2):91–110
55. Magnusson M, Lilienthal A, Duckett T (2007) Scan registration for autonomous mining vehicles using 3D-NDT. J Field Robot 24(10):803–827
56. Makadia A, Patterson IV A, Daniilidis K (2006) Fully automatic registration of 3D point clouds. In: Proceedings of the IEEE computer society conference on computer vision and pattern recognition, pp 1297–1304
57. Matas J, Chum O, Urban M, Pajdla T (2002) Robust wide baseline stereo from maximally stable extremal regions. In: Proceedings of the British machine vision conference, pp 36.1–36.10
58. Mikolajczyk K (2002) Detection of local features invariant to affine transformations. PhD thesis, Institut National Polytechnique de Grenoble, Grenoble, France
59. Minguez J, Montesano L, Lamiraux F (2006) Metric-based iterative closest point scan matching for sensor displacement estimation. IEEE Trans Robot 22(5):1047–1054
60. Moreno-Noguer F, Lepetit V, Fua P (2007) Accurate non-iterative $O(n)$ solution to the PnP problem. In: Proceedings of the IEEE international conference on computer vision, pp 1–8
61. Nistér D (2004) An efficient solution to the five-point relative pose problem. IEEE Trans Pattern Anal Mach Intell 26(6):756–770
62. Nüchter A, Gutev S, Borrmann D, Elseberg J (2011) Skyline-based registration of 3D laser scans. Geospat Inf Sci 14(2):85–90
63. Pathak K, Birk A, Vaškevičius N, Poppinga J (2010) Fast registration based on noisy planes with unknown correspondences for 3-D mapping. IEEE Trans Robot 26(3):424–441

64. Philip J (1998) Critical point configurations of the 5-, 6-, 7-, and 8-point algorithms for relative orientation. Technical Report TRITA-MAT-1998-MA-13, Department of Mathematics, Royal Institute of Technology, Stockholm, Sweden
65. Pizarro O, Eustice R, Singh H (2003) Relative pose estimation for instrumented, calibrated imaging platforms. In: Proceedings of digital image computing techniques and applications, pp 601–612
66. Pulli K, Piiroinen S, Duchamp T, Stuetzle W (2005) Projective surface matching of colored 3D scans. In: Proceedings of the international conference on 3-d digital imaging and modeling, pp 531–538
67. Rabbani T, Dijkman S, van den Heuvel F, Vosselman G (2007) An integrated approach for modelling and global registration of point clouds. ISPRS J Photogramm Remote Sens 61(6):355–370
68. Rodehorst V, Heinrichs M, Hellwich O (2008) Evaluation of relative pose estimation methods for multi-camera setups. Int Arch Photogramm Remote Sens Spat Inf Sci XXXVII-B3b:135–140
69. Rodrigues M, Fisher R, Liu Y (2002) Special issue on registration and fusion of range images. Comput Vis Image Underst 87(1–3):1–7
70. Rosten E, Drummond T (2005) Fusing points and lines for high performance tracking. In: Proceedings of the international conference on computer vision, vol 2, pp 1508–1515
71. Rublee E, Rabaud V, Konolige K, Bradski G (2011) ORB: an efficient alternative to SIFT or SURF. In: Proceedings of the IEEE international conference on computer vision, pp 2564–2571
72. Rusinkiewicz S, Levoy M (2001) Efficient variants of the ICP algorithm. In: Proceedings of the international conference on 3d digital imaging and modeling, pp 145–152
73. Rusu RB, Marton ZC, Blodow N, Beetz M (2008) Persistent point feature histograms for 3D point clouds. In: Proceedings of the international conference on intelligent autonomous systems, pp 119–128
74. Rusu RB, Blodow N, Beetz M (2009) Fast point feature histograms (FPFH) for 3D registration. In: Proceedings of the IEEE international conference on robotics and automation, pp 3212–3217
75. Salvi J, Matabosch C, Fofi D, Forest J (2007) A review of recent range image registration methods with accuracy evaluation. Image Vis Comput 25(5):578–596
76. Sattler T, Leibe B, Kobbelt L (2009) SCRAMSAC: improving RANSAC's efficiency with a spatial consistency filter. In: Proceedings of the IEEE international conference on computer vision, pp 2090–2097
77. Seo JK, Sharp GC, Lee SW (2005) Range data registration using photometric features. In: Proceedings of the IEEE computer society conference on computer vision and pattern recognition, vol 2, pp 1140–1145
78. Soudarissanane S, Lindenbergh R (2011) Optimizing terrestrial laser scanning measurement set-up. Int Arch Photogramm Remote Sens Spat Inf Sci XXXVIII-5/W12:127–132
79. Stamos I, Leordeanu M (2003) Automated feature-based range registration of urban scenes of large scale. In: Proceedings of the IEEE computer society conference on computer vision and pattern recognition, vol 2, pp 555–561
80. Steder B, Grisetti G, Van Loock M, Burgard W (2009) Robust on-line model-based object detection from range images. In: Proceedings of the IEEE/RSJ international conference on intelligent robots and systems, pp 4739–4744
81. Steder B, Grisetti G, Burgard W (2010) Robust place recognition for 3D range data based on point features. In: Proceedings of the IEEE international conference on robotics and automation, pp 1400–1405
82. Steder B, Rusu RB, Konolige K, Burgard W (2011) Point feature extraction on 3D range scans taking into account object boundaries. In: Proceedings of the IEEE international conference on robotics and automation, pp 2601–2608
83. Stewénius H, Engels C, Nistér D (2006) Recent developments on direct relative orientation. ISPRS J Photogramm Remote Sens 60(4):284–294

84. Tam GKL, Cheng ZQ, Lai YK, Langbein FC, Liu Y, Marshall D, Martin RR, Sun XF, Rosin PL (2013) Registration of 3D point clouds and meshes: a survey from rigid to nonrigid. IEEE Trans Vis Comput Graph 19(7):1199–1217

85. Theiler PW, Wegner JD, Schindler K (2012) Automatic registration of terrestrial laser scanner point clouds using natural planar surfaces. ISPRS Ann Photogramm Remote Sens Spat Inf Sci I-3:173–178

86. Theiler PW, Wegner JD, Schindler K (2013) Markerless point cloud registration with keypoint-based 4-points congruent sets. ISPRS Ann Photogramm Remote Sens Spat Inf Sci II-5/W2:283–288

87. Theiler PW, Wegner JD, Schindler K (2014) Fast registration of laser scans with 4-points congruent sets—What works and what doesn't. ISPRS Ann Photogramm Remote Sens Spat Inf Sci II-3:149–156

88. Theiler PW, Wegner JD, Schindler K (2014) Keypoint-based 4-points congruent sets—Automated marker-less registration of laser scans. ISPRS J Photogramm Remote Sens 96:149–163

89. Tola E, Lepetit V, Fua P (2008) A fast local descriptor for dense matching. In: Proceedings of the IEEE computer society conference on computer vision and pattern recognition, pp 1–8

90. Tola E, Lepetit V, Fua P (2010) DAISY: an efficient dense descriptor applied to wide-baseline stereo. IEEE Trans Pattern Anal Mach Intell 32(5):815–830

91. Trzcinski T, Christoudias M, Lepetit V, Fua P (2012) Learning image descriptors with the boosting-trick. In: Proceedings of the annual conference on neural information processing systems, vol 1, pp 269–277

92. Trzcinski T, Christoudias M, Fua P, Lepetit V (2013) Boosting binary keypoint descriptors. In: Proceedings of the IEEE computer society conference on computer vision and pattern recognition, pp 2874–2881

93. Tuytelaars T, Mikolajczyk K (2008) Local invariant feature detectors: a survey. Found Trends Comput Graph Vis 3(3):177–280

94. Umeyama S (1991) Least-squares estimation of transformation parameters between two point patterns. IEEE Trans Pattern Anal Mach Intell 13(4):376–380

95. Urban S, Weinmann M (2015) Finding a good feature detector-descriptor combination for the 2D keypoint-based registration of TLS point clouds. ISPRS Ann Photogramm Remote Sens Spat Inf Sci II-3/W5:121–128

96. von Hansen W (2006) Robust automatic marker-free registration of terrestrial scan data. Int Arch Photogramm Remote Sens Spat Inf Sci XXXVI-3:105–110

97. Wang Z, Brenner C (2008) Point based registration of terrestrial laser data using intensity and geometry features. Int Arch Photogramm Remote Sens Spat Inf Sci XXXVII-B5:583–589

98. Weickert J, Scharr H (2002) A scheme for coherence-enhancing diffusion filtering with optimized rotation invariance. J Vis Commun Image Represent 13(1–2):103–118

99. Weinmann M (2013) Visual features—From early concepts to modern computer vision. In: Farinella GM, Battiato S, Cipolla R (eds) Advanced topics in computer vision. Advances in computer vision and pattern recognition. Springer, London, pp 1–34

100. Weinmann M, Jutzi B (2011) Fully automatic image-based registration of unorganized TLS data. Int Arch Photogramm Remote Sens Spat Inf Sci XXXVIII-5/W12:55–60

101. Weinmann M, Jutzi B (2012) A step towards dynamic scene analysis with active multi-view range imaging systems. Int Arch Photogramm Remote Sens Spat Inf Sci XXXIX-B3:433–438

102. Weinmann M, Jutzi B (2013) Fast and accurate point cloud registration by exploiting inverse cumulative histograms (ICHs). In: Proceedings of the joint urban remote sensing event (JURSE), pp 218–221

103. Weinmann M, Jutzi B (2015) Geometric point quality assessment for the automated, marker-less and robust registration of unordered TLS point clouds. ISPRS Ann Photogramm Remote Sens Spat Inf Sci II-3/W5:89–96

104. Weinmann M, Weinmann Mi, Hinz S, Jutzi B (2011) Fast and automatic image-based registration of TLS data. ISPRS J Photogramm Remote Sens 66(6):S62–S70

105. Weinmann M, Wursthorn S, Jutzi B (2011) Semi-automatic image-based fusion of range imaging data with different characteristics. Int Arch Photogramm Remote Sens Spat Inf Sci XXXVIII-3/W22:119–124
106. Weinmann M, Dittrich A, Hinz S, Jutzi B (2013) Automatic feature-based point cloud registration for a moving sensor platform. Int Arch Photogramm Remote Sens Spat Inf Sci XL-1/W1:373–378
107. Wendt A (2007) A concept for feature based data registration by simultaneous consideration of laser scanner data and photogrammetric images. ISPRS J Photogramm Remote Sens 62(2):122–134
108. Wu Y, Hu Z (2006) PnP problem revisited. J Math Imaging Vis 24(1):131–141
109. Zeisl B, Köser K, Pollefeys M (2013) Automatic registration of RGB-D scans via salient directions. In: Proceedings of the IEEE international conference on computer vision, pp 2808–2815

Chapter 5
Co-Registration of 2D Imagery and 3D Point Cloud Data

While a geometric enrichment of 3D point cloud data may be achieved by aligning acquired 3D point clouds in a common coordinate frame (see Chap. 4), it is sometimes not sufficient to adequately describe the contents of a scene. When using a terrestrial/mobile laser scanner or a range camera, for instance, the acquired data encapsulating geometric information and intensity information allows an (almost) realistic depiction of a 3D scene which is typically of importance for tasks focusing on city modeling or cultural heritage applications. However, a variety of other applications focus on an automated analysis of the observed scene, where the complexity of typical real-world scenes often hinders important tasks such as object detection or scene interpretation. Consequently, the fusion of data acquired with complementary data sources is desirable as complementary information may be expected to facilitate a variety of tasks. Taking into account that both terrestrial/mobile laser scanners and range cameras allow to derive 2D representations in the form of range and intensity images (see Sect. 2.4), we may exploit 3D geometry and 2D imagery for data fusion when using a respective scanning device. Furthermore, we take into account that complementary data may, for instance, be acquired with digital cameras, thermal cameras, or multispectral cameras, where the acquired data is represented in the form of 2D images. Accordingly, robust feature correspondences have to be derived between the acquired data of different devices in order to allow a reliable 3D mapping of complementary 2D imagery onto the 3D point cloud data acquired with the scanning device.

In this chapter, we address the fact that particularly thermal information offers many advantages for scene analysis, since people may easily be detected as heat sources in typical indoor or outdoor environments and, furthermore, a variety of concealed objects such as heating pipes as well as structural properties such as defects in isolation may be observed. Additionally, a 3D mapping involving a range camera with high frame rate and a thermal camera with typical video frame rate may be helpful to describe the evolution of a dynamic 3D scene over time. In order to achieve a respective 3D mapping, we present a novel and fully automatic framework consisting of four successive components: (i) a radiometric correction, (ii) a geometric calibration, (iii) a robust approach for detecting reliable feature correspondences, and

© Springer International Publishing Switzerland 2016
M. Weinmann, *Reconstruction and Analysis of 3D Scenes*,
DOI 10.1007/978-3-319-29246-5_5

(iv) a co-registration of 3D point cloud data and thermal information. For the last component, we consider two different approaches represented by a RANSAC-based homography estimation for almost planar scenes and a RANSAC-based projective scan matching technique for general scenes. For the example of an indoor scene, we demonstrate the performance of our framework in terms of both accuracy and applicability. We additionally show that efficient straightforward techniques allow a sharpening of the blurry thermal infrared information or a categorization of the acquired data with respect to background, people, passive scene manipulation, and active scene manipulation. Our framework has successfully undergone peer review [49, 50].

After providing a brief motivation for thermal 3D mapping (Sect. 5.1), we reflect related work on the co-registration of 2D imagery and 3D point cloud data (Sect. 5.2) and present our framework for keypoint-based 3D mapping of thermal information (Sect. 5.3). Subsequently, we demonstrate the performance of our framework for the example of an indoor scene (Sect. 5.4) and discuss the derived results (Sect. 5.5). Finally, we provide concluding remarks and suggestions for future work (Sect. 5.6).

5.1   Motivation and Contributions

The automated acquisition, description, and interpretation of static and dynamic 3D scenes still represents a topic of major interest in photogrammetry, remote sensing, computer vision, and robotics. Due to the recent technological advancements, a variety of devices is currently available which may be used in order to acquire different types of information such as color, intensity, temperature, or spatial 3D geometry. Thus, using different devices mounted on a common sensor platform seems to be very promising as it allows to collect multidimensional spatial data. In particular, a sensor platform equipped with devices delivering complementary types of information may facilitate a variety of tasks and thus offer a high potential for numerous applications.

When designing a suitable sensor platform for 3D scene analysis, a scanning device (e.g., a laser scanner or a range camera) is required which provides spatial 3D information in the form of 3D point cloud data and radiometric information represented as intensity images in terms of either color or gray-valued images. As the complexity of real-world scenes typically hinders an adequate scene analysis based on only the data acquired with the scanning device, the acquisition of further, complementary information is desirable which may be achieved by involving other devices such as thermal cameras providing thermal information in the form of thermal infrared images. As the respective images reveal a significantly different behavior, a fusion of intensity information provided by the scanning device and thermal information provided by the thermal camera may clearly result in synergistic effects concerning scene analysis:

- Standard intensity images (i.e., color or gray-valued images) typically represent electromagnetic radiation in the visual domain and thus radiometric information influenced by surface properties of observed objects. The main influencing factors are represented by the illumination of the scene, the material of respective objects, the surface reflectivity, the surface roughness, and the relative geometric orientation between the respective surface and the utilized device.
- Thermal infrared images represent thermal radiation in the infrared spectrum. This radiation is emitted by objects in the scene and not visible in the visual domain. Consequently, thermal infrared images allow a different look on objects and the extraction of additional information like temperature and different materials of observed objects.

Accordingly, objects visible in the visual domain may be invisible in the thermal infrared domain if they have the same temperature and emissivity coefficient as the respective background. In contrast, in thermal infrared images, even further objects below the surface of an object may be visible which certainly remain invisible in the visual domain. Note that, in this context, two different materials with the same temperature may appear differently in thermal infrared images if they have a significantly different emissivity coefficient. Interestingly, two objects with different temperature and different emissivity coefficient may even coincidentally appear with high similarity in thermal infrared images. As a consequence, a fusion of standard intensity images and thermal infrared images may enhance specific features in these images or even reveal information which may not be observed in either intensity images or thermal infrared images [2, 10]. This is, for instance, of special interest for enhancing contrast in environments of poor visibility or inadequate illumination [28], for target detection [53] or for concealed object detection [52]. More importantly, infrared thermography facilitates building diagnostics [3] which, due to current attempts for saving energy, has become a research topic itself.

Whereas the fusion of intensity information provided by the scanning device and thermal information provided by the thermal camera exploits different types of information, it does not account for the respective spatial dimensions which, in turn, would allow a thermal 3D mapping in terms of projecting the thermal infrared image onto acquired 3D point cloud data. Such a thermal 3D mapping allows the quantification of thermal studies if the 3D point clouds are textured with thermal information [24]. Particularly in building observation [16, 17, 20], the joint representation of building façades and thermal information is desirable, since it allows a reconstruction of the surface temperature and with it a look into the interior behavior of a wall. Based on the surface temperature, valuable insights about different materials, heating pipes and leakages may easily be obtained and added as semantic or geometric information to a respective building model.

In order to capture co-registered intensity information, thermal information, and spatial 3D geometry, different sensor platforms have been presented for data acquisition within static and dynamic environments. Focusing on the acquisition of static scenes, the combination of data captured with a terrestrial laser scanner and images acquired with a bicamera system, i.e., a system consisting of an optical camera and a thermal camera, has recently been proposed in order to obtain 3D building models textured with respect to either color or thermal information [1]. Furthermore, a robot equipped with a laser scanner, a thermal camera, and a digital camera has been presented [6, 7], where the laser scanner generates a precise 3D model, the thermal camera reveals the heat distribution in the scene, and the additional color information may, for instance, be used to identify heat sources or to obtain photo-realistic 3D models. Focusing on the acquisition of dynamic scenes, respective systems addressing thermal 3D mapping involve range cameras such as RGB-D cameras or Time-of-Flight (ToF) cameras. For instance, the use of a low-cost RGB-D camera in combination with a thermal camera has recently been proposed in order to acquire dense 3D models of environments with both appearance and temperature information [45].

In this chapter, we focus on thermal 3D mapping which allows to observe the evolution of a dynamic 3D scene over time. Consequently, the scanning device involved for capturing spatial 3D information should be able to acquire scans in a small amount of time. In this regard, modern range cameras such as Microsoft Kinect, PMD[vision] CamCube 2.0 or MESA Imaging SR4000 simultaneously provide geometric and radiometric information in the form of range and intensity images with a relatively high frame rate of about 25 fps. Thus, they are also applicable in order to capture dynamic scenes. Accordingly, we consider the use of a platform equipped with a range camera and a thermal camera, both providing frame rates of about 25 fps. Our main contribution consists of a fully automatic framework for thermal 3D mapping which involves

- a geometric calibration of both the range camera and the thermal camera based on a common strategy,
- a robust approach for detecting 2D/2D feature correspondences via shape-based matching,
- a novel, but straightforward approach exploiting 2D/2D feature correspondences for the co-registration of 3D point cloud data and thermal information, and
- a novel projective scan matching technique exploiting 3D/2D correspondences for the co-registration of 3D point cloud data and thermal information.

While our approach based on 2D/2D feature correspondences is tailored to (almost) planar scenes as given when considering flat building façades or walls, our approach based on 3D/2D correspondences does not rely on such rather strong assumptions and may hence be applied for general scenes.

5.2 Related Work

For thermal 3D mapping involving a range camera, respective 3D coordinates are available for each pixel of the intensity image provided by the range camera. Accordingly, for thermal 3D mapping, feature correspondences have to be derived either between standard intensity images provided by the range camera (i.e., either color or gray-valued images) and thermal infrared images provided by the thermal camera

(Sect. 5.2.1), or between 3D point clouds provided by the range camera and thermal infrared images provided by the thermal camera (Sect. 5.2.2). Finally, for the sake of completeness, we also briefly reflect approaches for directly generating 3D point clouds from thermal infrared images (Sect. 5.2.3).

5.2.1 Indirect Co-Registration of 3D Point Clouds and Thermal Infrared Images

The first category of approaches for thermal 3D mapping focuses on indirectly acquiring textured 3D point cloud data. In this context, the 3D mapping is related to a standard *image registration* [55] in terms of estimating the respective transformation model between the considered images and subsequently warping the thermal infrared image provided by the thermal camera onto the intensity image provided by the range camera. Thus, a per-pixel assignment between spatial 3D information and respective thermal information is available which, in turn, allows to project the thermal information to 3D space by forward projection according to the respective 3D information.

Among a variety of approaches for image registration, the feature-based approaches are most widely spread. Thereby, the involved features are typically represented by visual features resulting from specific visual characteristics in the considered images. More specifically, according to our recent survey on visual features [46], different types of visual features may be categorized (see Sect. 3.2), where local features are the most prominent type due to their applicability for numerous applications such as image registration, image retrieval, object detection, object recognition, object tracking, navigation of autonomous vehicles, scene reconstruction, or scene interpretation. Respective approaches for extracting local features, however, are typically tailored to images acquired by either identical or similar devices from positions with a reasonable change in viewpoint. As a consequence, a matching between different image domains—e.g., a co-registration of satellite imagery and LiDAR intensity images [43]—may be quite challenging as the respective images reveal very different characteristics due to which many standard approaches for deriving feature correspondences tend to fail.

In particular, an automatic matching between images representing information in the visual domain and the thermal domain still remains challenging. Intensity images representing information in the visual domain typically provide sharp contours in the form of abrupt changes of properties like intensity or texture at the edges of objects in the observed scene. In thermal infrared images, however, we might face challenges arising from (i) the low geometric resolution compared to classical optical camera systems and (ii) the fact that features such as lines or contours do not exhibit sharp edges, but rather appear blurry. As a consequence, even powerful standard approaches such as the Scale Invariant Feature Transform (SIFT) [30, 31] are not applicable for automatically detecting feature correspondences between images representing information in the visual domain and the thermal domain.

For an automatic registration of image data representing information in these different spectral bands, respective approaches have been presented with a segment-based approach [11] or an approach involving normalized mutual information [36]. Considering mutual information between images has also been proposed for the mapping of multispectral texture information onto 3D models [38] and for the co-registration of intensity images and 3D LiDAR data [37]. Further approaches relying on a matching between images followed by a forward projection of 2D keypoints to 3D space may be based on the transformation model of a homography [14] once tie points have been detected. A respective approach has been proposed for mapping thermal information on existing building models [18].

5.2.2   Direct Co-Registration of 3D Point Clouds and Thermal Infrared Images

The second category of approaches for thermal 3D mapping focuses on directly acquiring textured 3D point cloud data. One of the simplest approaches in this regard consists of using co-registered devices with known fixed relative orientation [39]. If the relative orientation is unknown, the standard approach consists of manually selecting tie points and subsequently exploiting these tie points for a bundle adjustment based on the collinearity equations, which has recently been used in order to co-register 3D point clouds and thermal infrared images [19].

For a fully automatic estimation of the relative orientation between the used devices, silhouette-based approaches may be applied which focus on minimizing the error between the contour of an object in the image and the contour of the respective projected 3D model [29]. In particular, linear features are often used as they typically occur in man-made environments such as urban areas. For such scenes, lines may easily be extracted at edges in 2D imagery, and clusters of vertical and horizontal lines can be detected in the respective 3D point cloud. This allows a registration based on the matching of corresponding 2D and 3D linear features [26, 27].

5.2.3   Direct Generation of 3D Point Clouds from Thermal Infrared Images

In contrast to approaches involving scanning devices for thermal 3D mapping, it has also been proposed to directly generate 3D models from thermal infrared images. This may, for instance, be achieved with a thermal stereo system for recovering a 3D surface temperature map of the scene [40]. Interestingly, such a system even allows to estimate object depth within a dark environment. More specifically, this thermal stereo system is based on the idea of exploiting isotherms (i.e., lines connecting points of equal temperature) and epipolar geometry [14], whereby the epipolar constraints reduce the correspondence search space to the intersecting points between epipolar lines and isotherms. Besides this approach based on the strategy of stereo matching, it is also possible to establish thermal 3D mapping via Structure-from-Motion (SfM) techniques [34] which may generally be applied in order to simultaneously recover both the 3D structure of the scene and the respective pose (i.e., position and orientation) of the involved camera [42].

5.3 A Novel Framework for Keypoint-Based 3D Mapping of Thermal Information

For thermal 3D mapping, we propose a novel framework which automatically recovers the relative orientation between a range camera and a thermal camera. As shown in Fig. 5.1, the framework consists of four different components. After an initial radiometric correction (Sect. 5.3.1), we carry out a geometric calibration (Sect. 5.3.2) in order to obtain undistorted 2D imagery. Subsequently, we apply a shape-based matching technique in order to detect feature correspondences between intensity images provided by the range camera and thermal infrared images provided by the thermal camera (Sect. 5.3.3). Based on the derived feature correspondences, we present two approaches for the co-registration of 3D point cloud data and thermal information (Sect. 5.3.4): one of these approaches is tailored to (almost) planar scenes and has been presented in [49], whereas the other one is designed for general scenes and has been presented in [50].

5.3.1 Radiometric Correction

For standard range cameras such as PMD[vision] CamCube 2.0 or MESA Imaging SR4000, the captured radiometric information is represented in the form of either active intensity information (depending on the illumination emitted by the device) or passive intensity information (depending on the background illumination arising from the sun or other external light sources). Due to internal processes such as

Fig. 5.1 The proposed framework for the co-registration of 3D point cloud data provided by a range camera and 2D imagery provided by a thermal camera

Fig. 5.2 Visualization of data captured with a range camera of type PMD[vision] CamCube 2.0: active intensity image $\mathcal{I}_{I,a}$, passive intensity image $\mathcal{I}_{I,p}$, and range image $\mathcal{I}_R$ (from *left* to *right*). The range is encoded in terms of a color scale reaching from *red* (*near*) via *yellow*, *green*, *cyan*, and *blue* to *violet* (*far*)

the conversion to a digital signal and signal amplification (which are not identical for different types of range cameras), this intensity information has to be adapted in order to allow an application of standard image processing techniques. For this purpose, we apply a histogram normalization as described in Sect. 2.4 which adapts the intensity information to the interval [0, 255] and thus yields an 8-bit gray-valued image. Note that the histogram normalization may be carried out for both the acquired active intensity information and the acquired passive intensity information, and thus representations in the form of an active intensity image $\mathcal{I}_{I,a}$ and a passive intensity image $\mathcal{I}_{I,p}$ are available. For the example of an indoor scene, the derived range and intensity images acquired with a range camera of type PMD[vision] CamCube 2.0 are visualized in Fig. 5.2.

For thermal cameras, there is principally also a need for adapting the acquired thermal information to a visual 2D representation. However, this is typically already performed by internal software assigning each pixel a color value according to a certain colorbar and hence there is no need to apply a histogram normalization in this case. In order to obtain an impression about the behavior of acquired thermal information in the form of thermal infrared images $\mathcal{I}_T$, a respective visualization of thermal information captured with a thermal camera of type InfraTec VarioCAM hr is depicted in Fig. 5.3 for the example of an indoor scene.

5.3.2 Geometric Calibration

When considering Figs. 5.2 and 5.3, it clearly becomes visible that the acquired range and intensity images as well as the acquired thermal infrared images are distorted. For both devices, this distortion arises from the respective optical lens and hence the distortion may be described in analogy to the distortion caused by standard digital cameras. Accordingly, a *geometric calibration* has to be carried out for both the range camera and the thermal camera.

Fig. 5.3 Visualization of thermal information in the form of a thermal infrared image $\mathscr{I}_T$ captured with a thermal camera of type InfraTec VarioCAM hr

For the geometric calibration of range cameras, we may apply a standard camera calibration, where we assume that the geometric mapping of a scene onto the image plane may be described with a standard camera model (i.e., a generalized form of the pinhole camera model) representing the intrinsic behavior of a digital camera. This standard model considers both radial and tangential distortion [9, 15]. Accordingly, the geometric mapping follows the collinearity equations and may hence be parameterized with the focal lengths in x- and y-direction, the image coordinates (x_0, y_0) of the principal point, a skew coefficient s, and the image distortion coefficients describing radial and tangential distortion. The commonly used and well-known standard methodology for geometric calibration consists of using a rectangular checkerboard pattern with known size, capturing respective intensity images with the device, extracting the grid corners of the checkerboard pattern in the intensity images, and finally applying the calibration procedure [8, 54]. For our example depicted in Fig. 5.2, the respectively derived undistorted images are provided in Fig. 5.4. Based on the undistorted images, we may exploit the angular resolution of the scanning device and the respective range measurements for all pixels in order to derive the respective 3D information in terms of spatial 3D coordinates with respect to the local coordinate frame of the scanning device.

For the geometric calibration of the thermal camera, we have to take into account that the checkerboard pattern is not visible in the thermal infrared domain. Consequently, we try to imitate a pattern corresponding to the grid corners of a checkerboard pattern. For this purpose, we use a planar testfield with lamps [7, 16, 19, 23, 32].

Fig. 5.4 Visualization of the undistorted images for active intensity information, passive intensity information, and range information (from *left* to *right*)

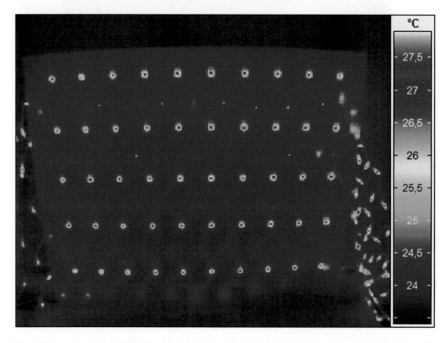

Fig. 5.5 Acquired thermal information for the planar testfield with lamps [16, 19]. Note that the distortion of the thermal infrared image is clearly visible

These lamps are clearly visible in the thermal infrared images as shown in Fig. 5.5 and may thus easily be detected. As a consequence, we may exploit the regular grid of lamps as input for the aforementioned standard calibration procedure for digital cameras which, in turn, yields undistorted thermal infrared images as shown in Fig. 5.6.

Fig. 5.6 Visualization of the undistorted thermal infrared image

5.3.3 Feature Extraction and Matching

Once undistorted images are available, the next step consists of deriving feature correspondences between the data acquired with the range camera and the thermal camera. For this purpose, we consider the derived image representations in the form of an intensity image $\mathscr{I}_I$ and a thermal infrared image $\mathscr{I}_T$, and we take into account that the least constraints with respect to estimating the relative orientation between both devices are given when using local features in terms of keypoints with respective descriptors.

In order to obtain reference values, we first carry out a manual extraction of feature correspondences. This involves a manual selection of points in the different images, an implicit description of the respective local characteristics and the detection of corresponding features by comparing the implicit descriptions. As human visual perception is extremely robust and supported by experience gained over a long time, a user may easily perform these steps and thus provide a set of reliable point correspondences, which we refer to as 2D/2D feature correspondences, without outliers (Fig. 5.7).

For reasons of efficiency, an automated feature extraction and matching would be desirable. Due to the different characteristics of the compared images, however, commonly used keypoint detectors and descriptors obtained from algorithms such as SIFT [30, 31] or SURF [4, 5] (see Sect. 4.3.3) typically fail in the automatic recovery of 2D/2D feature correspondences. Hence, we have to focus on different properties contained in the considered images. In this regard, exploiting shape information seems to be an interesting alternative, and a respective approach has been presented

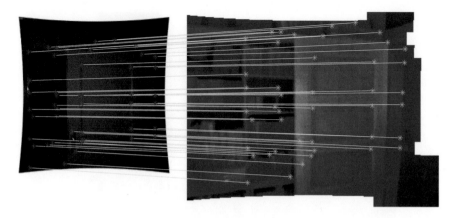

Fig. 5.7 Manually detected 2D/2D feature correspondences between an intensity image and a thermal infrared image for a static scene

with the *shape-based matching algorithm* [41, 44]. This algorithm is based on the idea of matching image patches of a given, user-defined size by exploiting wavelength-independent properties in terms of shape information. Thereby, the surrounding of a keypoint is described by a generated model. As a result, the shape-based matching algorithm still allows to robustly derive 2D/2D feature correspondences between the considered images, but a certain amount of outliers may still be expected due to repetitive patterns with respect to shape information.

More specifically, in our case, we first generate a model image by selecting quadratic areas of 100×100 pixels around given points on a discrete grid with a spacing of 10 pixels in the intensity image provided by the range camera. Here, the values for the size of the model image and the grid spacing have been selected empirically [49, 50]. Subsequently, a gradient filter in the form of a Sobel filter is applied to the model image (Fig. 5.8, left), and the associated gradient directions are determined

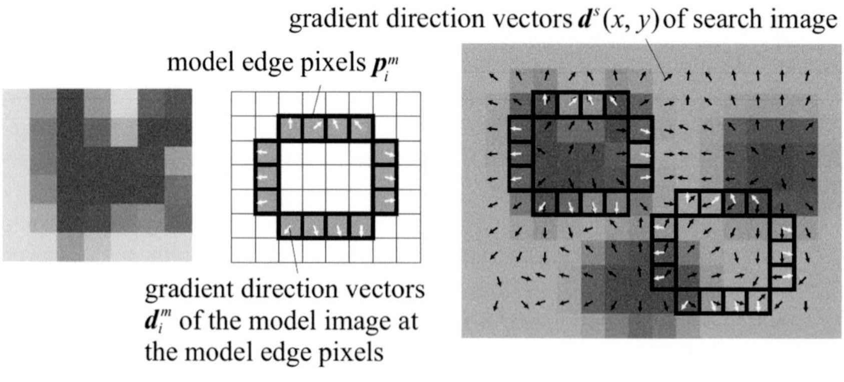

gradient direction vectors $\boldsymbol{d}^s(x, y)$ of search image

model edge pixels $\boldsymbol{p}_i^m$

gradient direction vectors
$\boldsymbol{d}_i^m$ of the model image at
the model edge pixels

Fig. 5.8 Principle of shape-based matching according to [44]: model image (*left*), model edges (*center*), and search image (*right*)

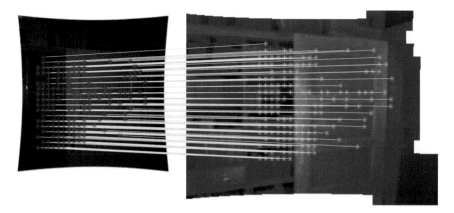

Fig. 5.9 Automatically detected 2D/2D feature correspondences between an intensity image and a thermal infrared image for a static scene

for those pixels with high gradient magnitude (Fig. 5.8, center). Finally, the model image is matched to the gradients of the search image—which is represented by the thermal infrared image—by comparing the respective gradient directions (Fig. 5.8, right). Thereby, a similarity measure is exploited which represents the normalized dot product of vector representations for the gradient directions of the transformed model and the search image [44], according to which a score may be obtained for each pixel in the search image. According to [41, 44], this similarity measure is robust in case of noise, changes in illumination and partial occlusions, but not in case of changes in rotation and scale. Hence, the search space is extended to a predefined range of rotations and scales. If the derived similarity measure is above a certain threshold, a 2D/2D feature correspondence is detected (Fig. 5.9). For each 2D/2D feature correspondence, the coordinates of the respective point on the discrete grid of the intensity image provided by the range camera are stored as well as the respective coordinates of the center of the best-matching model image in the thermal infrared image, the respective rotation angle, and the calculated similarity measure itself. In our experiments, we use the HALCON 11 implementation (MVTec Software) of the shape-based matching algorithm.

5.3.4 Keypoint-Based Co-Registration of 3D and 2D Information

Based on the 2D/2D feature correspondences derived between the intensity image provided by the range camera and the thermal infrared image provided by the thermal camera, it is possible to estimate the relative orientation between both devices. In the scope of this chapter, we consider two different strategies:

- The first strategy presented in Sect. 5.3.4.1 is based on the assumption of an (almost) planar scene and consists of a fundamental concept—the robust estimation of a homography—which is commonly applied for image registration. Based on the estimated transformation, the thermal infrared image provided by the thermal camera may easily be warped onto the intensity image provided by the range camera (thus a per-pixel assignment between thermal and range information is available which allows a forward projection of the thermal information to 3D space), and the estimated homography may be decomposed into a rotation matrix and a translation vector which, in turn, represent the relative orientation between both devices.
- The second strategy presented in Sect. 5.3.4.2 does not make assumptions on the structure of the scene and is therefore suitable for general scenes. Basically, this strategy consists of adapting the robust projective scan matching technique presented in Chap. 4 to the given data, and the relative orientation between both devices is derived by solving the Perspective-n-Point (PnP) problem.

5.3.4.1   Co-Registration via the Robust Estimation of a Homography

Once 2D/2D feature correspondences have been derived, these may be exploited in order to directly estimate the transformation between the respective images. For this purpose, different types of transformations may be taken into account such as a homography for planar scenes or a fundamental matrix for nonplanar scenes. Considering the application of gaining infrared-textured models of building façades in outdoor environments or infrared-textured models of walls in indoor environments, an almost planar scene is given for the relevant image regions. Hence, exploiting the transformation model of a homography is appropriate in order to estimate the transformation between the respective images and derive a superposition of these images.

Generally, a *homography* is a perspective transformation model for planes and it represents a linear transformation in the projective space $\mathbb{P}^2$, where 2D points $\mathbf{x}_1$ of a plane $\boldsymbol{\pi}_1$ are mapped onto 2D points $\mathbf{x}_2$ of a plane $\boldsymbol{\pi}_2$ according to

$$\mathbf{x}_2 = \mathbf{H} \cdot \mathbf{x}_1 \tag{5.1}$$

where the 2D points $\mathbf{x}_1$ and $\mathbf{x}_2$ are considered in homogeneous coordinates and $\mathbf{H} \in \mathbb{R}^{3 \times 3}$ represents the homography matrix. For applications involving camera-like devices, the planes $\boldsymbol{\pi}_1$ and $\boldsymbol{\pi}_2$ represent the respective image planes. A limitation regarding this projection model of a homography however may be identified in the fact that the respective observed 3D scene points $\mathbf{X}$ must not represent arbitrary points in 3D space, but only 3D points on a plane $\boldsymbol{\pi}$ as shown in Fig. 5.10.

The estimation of a homography matrix $\mathbf{H}$ is typically carried out via the normalized direct linear transformation based on $n \geq 4$ feature correspondences, and we may decompose the estimated homography (which is determined up to a scalar factor) according to

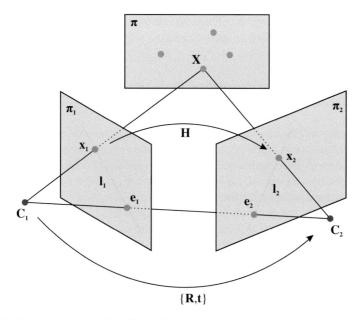

Fig. 5.10 Epipolar geometry, where $\mathbf{C}_1$ and $\mathbf{C}_2$ represent the projection centers, $\mathbf{e}_1$ and $\mathbf{e}_2$ represent the epipoles, and $\mathbf{l}_1$ and $\mathbf{l}_2$ represent the epipolar lines. The homography $\mathbf{H}$ describes a mapping of 2D points $\mathbf{x}_1$ of a plane π_1 onto 2D points $\mathbf{x}_2$ of a plane π_2 and may be decomposed to estimate the relative orientation $\{\mathbf{R}, \mathbf{t}\}$ between the different views

$$\mathbf{H} = \mathbf{R} + \frac{1}{d}\mathbf{tn}^T \qquad (5.2)$$

in order to recover the relative orientation between the involved devices by deriving the respective rotation matrix $\mathbf{R} \in \mathbb{R}^{3 \times 3}$ and the respective translation vector $\mathbf{t} \in \mathbb{R}^3$. As remaining symbols in this equation, $d \in \mathbb{R}$ represents the orthogonal distance to the planar scene and $\mathbf{n} \in \mathbb{R}^3$ represents the normal vector with respect to the planar scene. The homography decomposition yields four solutions for $\{\mathbf{R}, \mathbf{t}\}$. Of these, only two are valid when evaluating the positive depth constraint since the scene is of course in front of both devices and, by assuming that the (almost) planar scene is almost perpendicular to the viewing direction, we may determine the suitable solution as the solution corresponding to the normal vector most similar to the viewing direction. For more details, we refer to well-known standard literature on this topic [14, 33]. Since we may expect the presence of outliers, a robust estimation of the homography matrix is advisable for which we involve the RANSAC algorithm [13].

The estimated homography allows to warp the thermal infrared image provided by the thermal camera onto the intensity image provided by the range camera and, since the transformed image coordinates are determined with subpixel accuracy, a bilinear interpolation may be carried out in order to derive the respective information for those points on the 2D grid defined by the range camera. Thus, a per-pixel assignment between spatial 3D information and interpolated thermal information is available and, consequently, the interpolated thermal information may simply be projected to 3D space by forward projection according to the respective spatial 3D information.

5.3.4.2 Co-Registration via a Robust Projective Scan Matching Technique

For nonplanar and thus more general scenes, we may consider the estimation of a *fundamental matrix* in order to estimate the transformation between the respective images. However, the fundamental matrix only relates 2D points $\mathbf{x}_1$ of a plane π_1 with a respective epipolar line $\mathbf{l}_2$ in another plane π_2 and 2D points $\mathbf{x}_2$ of a plane π_2 with a respective epipolar line $\mathbf{l}_1$ in the plane π_1 as shown in Fig. 5.10. Thus, additional effort is, for instance, required to compare local image patches along the epipolar line $\mathbf{l}_2$ in π_2 with the local image patch around $\mathbf{x}_1$ in π_1, and a high number of comparisons would be necessary if we want to transform each pixel on the image raster corresponding to the plane π_1 to the image raster corresponding to the plane π_2. Furthermore, we may take into account that we already have 3D information acquired with the range camera. Consequently, the method of choice consists of deriving 3D/2D correspondences instead of 2D/2D feature correspondences and transferring the task of co-registration to projective scan matching presented in Sect. 4.3.6.1.

Since the range camera provides range and intensity information for the same grid, we may directly assign each of the derived 2D/2D feature correspondences a respective 3D scene point whose coordinates are known in the local coordinate frame of the range camera. Thus, we may easily define 3D/2D correspondences, where the 3D information is derived via the range camera and the 2D information is derived via the thermal camera. Depending on a variety of influencing factors (see Sect. 2.5.1), the range camera provides more or less noisy range information. Note that many points which arise from objects in the scene will probably provide a smooth surface. However, range information of points along edges of the respective objects might be very noisy. In order to detect such noisy measurements, we apply our approach presented in Sect. 2.5.3 which is based on the measure $\sigma_{r,3\times3}$ of range reliability defined as the standard deviation of all range values within the (3×3) image neighborhood of a pixel [47]. This measure is calculated for each point on the regular 2D grid and used as a measure describing the reliability of the respective range information. Accordingly, we are able to derive a continuous 2D confidence map which is illustrated in Fig. 5.11 for two exemplary scenes. Based on the derived 2D confidence map, a simple thresholding is sufficient to distinguish between reliable and unreliable range measurements. If, for a considered pixel (x, y), the measure $\sigma_{r,3\times3}$ of range reliability is larger than a predefined threshold t_σ, the respective range measurement is assumed to be unreliable, otherwise it is assumed to be reliable. Transferred to the considered 2D/2D feature correspondences, an unreliable correspondence is detected if a value of $\sigma_{r,3\times3} > t_\sigma$ is given for at least one of the respective image locations. Following our investigations in [48], we select a threshold of $t_\sigma = 0.05\,\text{m}$ in order to obtain reliable 3D/2D correspondences.

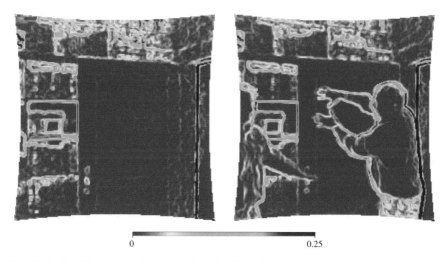

0   0.25

Fig. 5.11 Confidence maps indicating the reliability of the respective range information. The reliability is encoded in terms of a color scale reaching from *red* (reliable) via *yellow*, *green*, *cyan*, and *blue* to *violet* (unreliable)

Once 3D/2D correspondences have been derived, the task of co-registering 3D and 2D information may be solved via projective scan matching as explained in Sect. 4.3.6.1, i.e., by relating this task to solving the well-known Perspective-n-Point (PnP) problem, where the aim is to estimate the exterior orientation or pose of a camera from a set of n correspondences between 3D points $\mathbf{X}_i$ with $i = 1, \ldots, n$ of a scene and their 2D projections $\mathbf{x}_i$ onto the image plane [13, 51]. In order to solve the PnP problem, we again make use of the Efficient Perspective-n-Point (EPnP) algorithm [25, 35] which represents a noniterative method and provides an accurate solution to the PnP problem with only linear complexity. Compared to other approaches for solving the PnP problem, this algorithm is not only fast and accurate, but also designed to work with a large number of correspondences and it does not require an initial estimate for the transformation parameters.

For increased robustness in case of existing outlier correspondences, we again involve the RANSAC algorithm [13] which eliminates the influence of outlier correspondences not being in accordance with the largest consensus set. Following the original implementation [35], the RANSAC-based EPnP scheme relies on selecting small, but not minimal subsets of seven correspondences in order to estimate the model parameters and checking the whole set of correspondences for consistent samples. This further reduces the sensitivity to noise in comparison to minimal subsets. In order to avoid testing all possible subsets, which would be very time-consuming, we exploit an efficient variant, where the number of iterations—which equals the number of randomly chosen subsets—is selected sufficiently high, so that a subset including only inlier correspondences is selected with a certain probability p [13, 14].

Without loss of generality, we may assume that the local coordinate frame of the range camera represents the reference coordinate frame. Consequently, the derived 3D coordinates are known with respect to the reference frame. Together with the respective observed 2D image locations in the thermal infrared image, they form the required 3D/2D correspondences. Note that, for this reason, the calibration matrix **K** in Eq. (4.18) refers to the thermal camera in case of thermal 3D mapping.

5.4   Experimental Results

In the following, we first describe the main characteristics of our system used for data acquisition (Sect. 5.4.1). Subsequently, we outline the conducted experiments (Sect. 5.4.2) and present the respective results (Sect. 5.4.3).

5.4.1   Data Acquisition

Since our system should generally be suited for thermal 3D mapping within both indoor and outdoor scenes, the use of scanning devices relying on structured light projection (e.g., a Microsoft Kinect) is impractical due to larger distances between the scanning device and objects in the scene, but also due to external influences such as sunlight. Furthermore, we also intend to observe dynamic scenes where terrestrial/mobile laser scanners fail as they typically perform a successive spatial scanning of the observed scene by successively capturing spatial 3D coordinates corresponding to points on a discrete scan grid. As a consequence, for thermal 3D mapping, we use a range camera of type PMD[vision] CamCube 2.0 and a thermal camera of type InfraTec VarioCAM hr. Both devices are mounted on a sensor platform as shown in Fig. 5.12, so that a fixed relative orientation between the two devices is preserved. More technical details on the involved sensors are provided in the following subsections.

5.4.1.1   Range Camera—PMD[Vision] CamCube 2.0

The involved range camera is a PMD[vision] CamCube 2.0 which provides the capability to record both geometric and radiometric information—and thus complementary types of data—in the form of images with a single shot, i.e., the information is acquired simultaneously for all points on the discrete, regular 2D grid. For each point on this discrete grid and thus for each pixel of the respective image representation, three features are measured: range, active intensity, and passive intensity. In this context, the active intensity depends on the illumination emitted by the sensor, whereas the passive intensity depends on the background illumination arising from the sun or other external light sources. The acquired images have a size of 204×204 pixels

Fig. 5.12 Sensor platform equipped with a range camera of type PMD[vision] CamCube 2.0 (*left*) and a thermal camera of type InfraTec VarioCAM hr (*right*)

which corresponds to a field-of-view of $40° \times 40°$ and, thus, the device provides measurements with an angular resolution of approximately $0.2°$. As a result, a single frame $\mathscr{F}$ consists of a range image $\mathscr{I}_R$, an active intensity image $\mathscr{I}_{I,a}$ and a passive intensity image $\mathscr{I}_{I,p}$, and all measurements can be updated with high frame rates of more than 25 releases per second. Consequently, such a range camera also allows to capture dynamic scenes.

Due to the exploited measurement principle, the nonambiguous range depends on the modulation frequency. A modulation frequency of 20 MHz, for instance, results in a nonambiguous range of 7.5 m. In order to overcome such a range measurement restriction, image- or hardware-based unwrapping procedures have recently been introduced [12, 21, 22].

5.4.1.2 Thermal Camera—InfraTec VarioCAM Hr

The involved thermal camera is a bolometer-based InfraTec VarioCAM hr. It records electromagnetic radiation in the wavelength interval between 7.5 and 14 μm (i.e., in the thermal infrared domain) with a radiometric resolution of 0.05 K. The acquired thermal information is represented in the form of a thermal infrared image with a size of 384×288 pixels. The angular resolution of the device is approximately $0.16°$ and, thus, an acquired thermal infrared image corresponds to a field-of-view of about $61° \times 46°$. With a given frame rate of 25 fps, this device may also be applied for observing dynamic scenes.

5.4.2   Experiments

In order to test the performance of our framework, we only have to estimate the relative orientation between the involved devices once, which is done for the example of a static scene with the two presented strategies represented by a RANSAC-based homography estimation for almost planar scenes and a RANSAC-based projective scan matching technique for general scenes. Afterward, co-registered data may be recorded continuously when using a (hardware or software) trigger for synchronization, since the relative orientation between the utilized devices is known and remains fixed.

5.4.3   Results

Using the sensor platform shown in Fig. 5.12, we first observe a static and almost planar indoor scene and apply our framework for thermal 3D mapping. Subsequently, we consider the same scene after people have entered the scene and thus violate the assumption of an almost planar scene.

First, we carry out a manual extraction of 2D/2D feature correspondences. We select 44 feature correspondences as illustrated in Fig. 5.7, and since human perception is relatively robust in case of thoroughly selecting corresponding points, we may consider all these 2D/2D feature correspondences as inliers. The results for image registration relying on the estimation of a homography are shown in Fig. 5.13 (left). The respective 3D mapping based on the transformation model of a homography serves as ground truth representing the state-of-the-art accuracy.

Fig. 5.13 Passive intensity image of the range camera and transformed thermal infrared image when applying a homography estimation relying on a manual feature matching (*left*) and a RANSAC-based homography estimation relying on an automatic feature matching (*right*). The results may hardly be distinguished

Exploiting the shape-based matching algorithm in order to automatically derive 2D/2D feature correspondences, we obtain a total number of 151 feature correspondences as shown in Fig. 5.9. Of these, 14 feature correspondences are discarded since the respective range information is considered to be unreliable according to the applied rejection strategy based on the measure $\sigma_{r,3 \times 3}$ of range reliability (Fig. 5.11, left), for which a threshold of $t_\sigma = 0.05$ m is applied in order to distinguish between reliable and unreliable 2D/2D feature correspondences. The remaining 137 feature correspondences are exploited by a RANSAC-based homography estimation for which the image registration results and the respective results of thermal 3D mapping are shown in Figs. 5.13 (right) and 5.14 (left). Instead of 2D/2D feature correspondences, the respective 3D/2D correspondences are exploited by the RANSAC-based projective scan matching technique. The respective results of thermal 3D mapping are shown in Fig. 5.15 (left). Once the relative orientation is known, we may conduct thermal 3D mapping by exploiting the known transformation parameters, e.g., as shown in Figs. 5.14 (right) and 5.15 (right).

5.5 Discussion

The results derived with the RANSAC-based homography estimation (Fig. 5.14) reveal that this strategy is well suited for thermal 3D mapping in case of almost planar scenes. The basic constraints to be satisfied have been identified in a robust wavelength-independent mapping between image data with different characteristics, where standard approaches typically tend to fail. In this regard, the shape-based matching algorithm still allows to reliably derive 2D/2D feature correspondences, and a respective image registration based on these 2D/2D feature correspondences yields very similar results as an image registration based on manually selected 2D/2D feature correspondences (Fig. 5.13). The accurate image registration, in turn, is essential for a subsequent forward projection of the transformed thermal infrared information to 3D space according to the respective range information. However, a RANSAC-based homography estimation is based on the assumption of (almost) planar scenes, i.e., the scene depth is much smaller than the distance between the sensor platform and the scene. As a consequence, objects with a larger distance to the respective plane are not appropriately represented in the 3D point cloud with the expected thermal information (Fig. 5.14, right).

The results derived with the RANSAC-based projective scan matching technique (Fig. 5.15) clearly reveal that this strategy is well suited for thermal 3D mapping. In particular, the quality is rather high due to a robust wavelength-independent mapping between image data with different characteristics, where standard approaches typically tend to fail, and due to the consideration of a transformation model which is suited for general scenes. Note that only a few pixels at the edges between the person to the right and the background may be identified as erroneous mapping (Fig. 5.15, right). In contrast to a RANSAC-based homography estimation (Fig. 5.14, right), the assignment of the respective thermal 3D information is much better for the person to the right.

Fig. 5.14 Visualization of the results for thermal 3D mapping by applying a RANSAC-based homography estimation: front views (*top*) and colored 3D point clouds (*center* and *bottom*)

Fig. 5.15 Visualization of the results for thermal 3D mapping by applying the RANSAC-based projective scan matching technique: 2D image projections (*top*) and colored 3D point clouds (*center* and *bottom*)

As a result, we recommend the second strategy to perform an appropriate thermal 3D mapping in dynamic scenes fully automatically without making strong assumptions on the 3D scene structure and without human interaction for selecting corresponding features. Thus, our RANSAC-based projective scan matching technique outperforms other recent approaches which, in turn, reveal limitations as they either rely on human interaction [19] or as they are only tailored to planar scene structures. Note that—without loss of generality—our framework could also be applied for co-registering 3D point cloud data and color information, where the latter is acquired with a digital camera. This may even be significantly less challenging since the contours in the intensity image of a range camera and in the color image of a digital camera are not as blurry as in thermal infrared images.

Image Sharpening As explained in Sect. 5.3.4.1, the estimated homography allows to warp the thermal infrared image $\mathscr{I}_T$ provided by the thermal camera onto the intensity image $\mathscr{I}_{I,p}$ provided by the range camera. Besides the fact that data captured with complementary data sources has been fused, it becomes obvious that the passive intensity image $\mathscr{I}_{I,p}$ provides relatively sharp texture information with high frequency components, whereas the thermal infrared image $\mathscr{I}_T$ only contains relatively blurry texture information with significantly lower frequency components. Hence, we may further establish synergistic effects arising from the different types of data. In this regard, we focus on image manipulation in terms of sharpening the co-registered thermal infrared information (i.e., the transformed thermal infrared image) by exploiting the respective intensity information captured with the range camera.

For a respective sharpening of the transformed thermal infrared image $\mathscr{T}(\mathscr{I}_T)$ (where the operator $\mathscr{T}(\cdot)$ indicates the transformation), the image regions to be affected by a possible image enhancement are typically located around edges in the intensity image $\mathscr{I}_{I,p}$. Hence, we first derive the edge image $\mathscr{E}(\mathscr{I}_{I,p})$ by convolving the intensity image $\mathscr{I}_{I,p}$ with a Laplacian filter $\mathbf{L}$ given as

$$\mathbf{L} = \begin{bmatrix} 0 & -1 & 0 \\ -1 & 4 & -1 \\ 0 & -1 & 0 \end{bmatrix} \tag{5.3}$$

according to

$$\mathscr{E}(\mathscr{I}_{I,p}) = \mathbf{L} * \mathscr{I}_{I,p}. \tag{5.4}$$

Subsequently, we may carry out the image sharpening by simply adding the weighted edge image to each channel of the transformed thermal infrared image $\mathscr{T}(\mathscr{I}_T)$ in the RGB color space according to

$$\mathscr{S}(\mathscr{T}(\mathscr{I}_T)) = \mathscr{T}(\mathscr{I}_T) + \beta \cdot \mathscr{E}(\mathscr{I}_{I,p}) \tag{5.5}$$

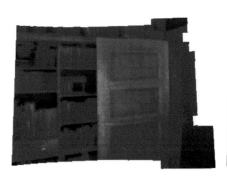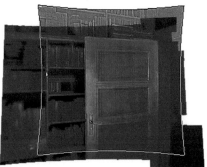

Fig. 5.16 Transformed thermal infrared image before sharpening (*left*) and after sharpening with $\beta = 1.0$ (*right*)

where $\beta \in \mathbb{R}$ is a constant parameter and the operator $\mathscr{S}(\cdot)$ indicates the image sharpening. Exemplary results before and after image sharpening based on a parameter of $\beta = 1.0$ are shown in Fig. 5.16 and clearly indicate the feasibility of the proposed approach. A forward projection of this sharpened thermal information onto the respective 3D point cloud yields enhanced thermal infrared-textured 3D models as shown in Fig. 5.17. Such manipulated 3D models, in turn, allow a more reliable extraction of features or primitives such as doors and windows in case of building diagnostics.

Scene Interpretation As explained in Sect. 5.3.4.2, thermal 3D mapping based on projective scan matching is not restricted to (almost) planar scenes and thus applicable for typical 3D scenes. In order to demonstrate the great potential arising from the accurately co-registered data, we focus on exploiting synergistic effects for scene interpretation. For this, we consider a respective thermal 3D mapping to observe the evolution of a 3D scene over time. Considering the two scenes depicted in Fig. 5.15 (left and right), the respective projections onto the image plane of the range camera (Fig. 5.15, top) and further involving the undistorted intensity and range images acquired with the range camera (e.g., Fig. 5.4) allows a categorization of the acquired data with respect to (i) *background* where no significant changes with respect to intensity, range or thermal information are observed, (ii) *people* in the scene which may be observed from a change in intensity, range, and thermal information, (iii) *passive scene manipulation* which is indicated only by a change in intensity information, and (iv) *active scene manipulation* caused by interactions between people and scene objects which is indicated only by a change in thermal information. Note that already the exploitation of thresholded difference images and a connection of logical operations is sufficient to derive a respective statement on change detection (Fig. 5.18). In the respective *rule-based classification* results according to the four considered classes of changes (Fig. 5.19), only a small error in the co-registration may be observed which is also visible in Fig. 5.15 (right). Accordingly, a respective segmentation of people in the 3D point cloud becomes trivial.

Fig. 5.17 Visualization of sharpened thermal information mapped onto the respective 3D point cloud for an almost planar scene (*top*) and a nonplanar scene (*bottom*)

Fig. 5.18 Thresholded difference images for intensity, range and thermal information (from *left* to *right*). Changes are indicated in *white*. Note the noisy behavior of range measurements (*center*)

Fig. 5.19 Results of a
rule-based classification
based on logical operators: a
clear distinction between
background (*blue*), *people*
(*red*), *passive scene
manipulation* (*yellow*) and
active scene manipulation
(*green*) is possible. Note that
passive and active scene
manipulation are also
detected at edges which
indicates a small error in the
co-registration

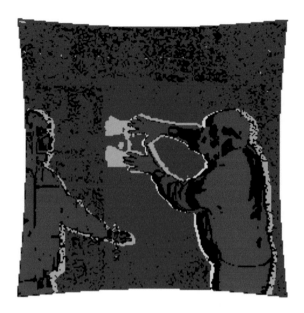

5.6 Conclusions

In this chapter, we have focused on thermal 3D mapping as an important prerequisite for building diagnostics in static scenes and object detection in dynamic scenes. Exploiting a sensor platform equipped with a range camera and a thermal camera, we have presented a fully automated framework for thermal 3D mapping which involves (i) a radiometric correction, (ii) a geometric calibration, (iii) a robust approach for detecting reliable feature correspondences between different image domains by exploiting wavelength-independent properties, and (iv) a co-registration of 3D point cloud data and thermal information based on two different strategies. While the first strategy for co-registration focuses on (almost) planar scenes, the second strategy for co-registration is well suited for general 3D scenes and thus outperforms other recent approaches in terms of accuracy and also in terms of applicability, due to avoiding both human interaction and strong assumptions on the 3D scene structure.

In order to show the great potential of thermal 3D mapping, we have presented two scenarios exploiting an appropriate co-registration of data acquired with a range camera and data acquired with a thermal camera. The first scenario addresses image sharpening in terms of sharpening the edges in the blurry thermal infrared images by exploiting the high quality of intensity images acquired with the range camera and thus allows to gain 3D models with enhanced thermal information. The second scenario addresses scene interpretation in terms of observing the evolution of a 3D scene over time which may easily be carried out via a simple thresholding technique followed by a rule-based classification.

For future research, it is possible to extend the presented framework with components focusing on 3D scene analysis. In this regard, particularly the detection of people and an automatic interpretation of their activities would be interesting, but also the detection of anomalies in both human behavior and local activity patterns. Furthermore, it would be desirable to not only observe a static or dynamic scene with devices mounted on a static sensor platform, but also to move the sensor platform in order to capture 3D environments of larger scale and extract complete 3D models which could be valuable for building diagnostics in both indoor and outdoor environments. In case of such a moving sensor platform, a further extension of our framework may be represented by automatically estimating the motion of the sensor platform and performing an automatic tracking of observed dynamic objects. Besides extensions of the framework, it would also be possible to integrate additional devices such as a standard digital camera or a multispectral camera in our acquisition system in order to gain further complementary information which could alleviate 3D scene analysis.

References

1. Alba MI, Barazzetti L, Scaioni M, Rosina E, Previtali M (2011) Mapping infrared data on terrestrial laser scanning 3D models of buildings. Remote Sens 3(9):1847–1870
2. Bai X, Zhou F, Xue B (2011) Fusion of infrared and visual images through region extraction by using multi scale center-surround top-hat transform. Opt Express 19(9):8444–8457
3. Balaras CA, Argiriou AA (2002) Infrared thermography for building diagnostics. Energy Build 34(2):171–183
4. Bay H, Tuytelaars T, Van Gool L (2006) SURF: speeded up robust features. In: Proceedings of the European conference on computer vision, vol 1, pp 404–417
5. Bay H, Ess A, Tuytelaars T, Van Gool L (2008) Speeded-up robust features (SURF). Comput Vis Image Underst 110(3):346–359
6. Borrmann D, Afzal H, Elseberg J, Nüchter A (2012) Mutual calibration for 3D thermal mapping. In: Proceedings of the international IFAC symposium on robot control, pp 605–610
7. Borrmann D, Elseberg J, Nüchter A (2012) Thermal 3D mapping of building façades. In: Proceedings of the international conference on intelligent autonomous systems, pp 173–182
8. Bouguet J-Y (2010) Camera calibration toolbox for Matlab. Computer Vision Research Group, Department of Electrical Engineering, California Institute of Technology, Pasadena, USA. http://www.vision.caltech.edu/bouguetj/calib_doc/index.html. Accessed 30 May 2015
9. Brown DC (1971) Close-range camera calibration. Photogramm Eng 37(8):855–866
10. Chen S, Leung H (2009) An EM-CI based approach to fusion of IR and visual images. In: Proceedings of the international conference on information fusion, pp 1325–1330
11. Coiras E, Santamaría J, Miravet C (2000) Segment-based registration technique for visual-IR images. Opt Eng 39(1):282–289
12. Droeschel D, Holz D, Behnke S (2010) Multi-frequency phase unwrapping for time-of-flight cameras. In: Proceedings of the IEEE/RSJ international conference on intelligent robots and systems, pp 1463–1469
13. Fischler MA, Bolles RC (1981) Random sample consensus: a paradigm for model fitting with applications to image analysis and automated cartography. Commun ACM 24(6):381–395
14. Hartley RI, Zisserman A (2008) Multiple view geometry in computer vision. University Press, Cambridge

15. Heikkilä J, Silvén O (1997) A four-step camera calibration procedure with implicit image correction. In: Proceedings of the IEEE computer society conference on computer vision and pattern recognition, pp 1106–1112
16. Hoegner L (2014) Automatische Texturierung von Fassaden aus terrestrischen Infrarot-Bildsequenzen. PhD thesis, Institut für Photogrammetrie und Kartographie, Technische Universität München, München, Germany
17. Hoegner L, Kumke H, Meng L, Stilla U (2007) Automatic extraction of textures from infrared image sequences and database integration for 3D building models. PFG—Photogramm Fernerkund Geoinf 6(2007):459–468
18. Hoegner L, Kumke H, Schwarz A, Meng L, Stilla U (2007) Strategies for texturing building models with low resolution infrared image sequences. Int Arch Photogramm Remote Sens Spat Inf Sci XXXVI-5/C55:1–6
19. Hoegner L, Roth L, Weinmann M, Jutzi B, Hinz S, Stilla U (2014) Fusion von Time-of-Flight-Entfernungsdaten und thermalen IR-Bildern. AVN—Allg Vermess-Nachr 5(2014):192–197
20. Iwaszczuk D, Hoegner L, Stilla U (2011) Detection of windows in IR building textures using masked correlation. In: Stilla U, Rottensteiner F, Mayer H, Jutzi B, Butenuth M (eds) Photogrammetric image analysis, ISPRS Conference—Proceedings. Lecture notes in computer science, vol 6952, Springer, Heidelberg, pp 133–146
21. Jutzi B (2009) Investigations on ambiguity unwrapping of range images. Int Arch Photogramm Remote Sens Spat Inf Sci XXXVIII-3/W8:265–270
22. Jutzi B (2012) Extending the range measurement capabilities of modulated range imaging devices by time-frequency-multiplexing. AVN—Allg Vermess-Nachr 2(2012):54–62
23. Lagüela S, González-Jorge H, Armesto J, Arias P (2011) Calibration and verification of thermographic cameras for geometric measurements. Infrared Phys Technol 54(2):92–99
24. Lagüela S, Martínez J, Armesto J, Arias P (2011) Energy efficiency studies through 3D laser scanning and thermographic technologies. Energy Build 43(6):1216–1221
25. Lepetit V, Moreno-Noguer F, Fua P (2009) EPnP: an accurate O(n) solution to the PnP problem. Int J Comput Vis 81(2):155–166
26. Liu L, Stamos I (2005) Automatic 3D to 2D registration for the photorealistic rendering of urban scenes. In: Proceedings of the IEEE computer society conference on computer vision and pattern recognition, vol 2, pp 137–143
27. Liu L, Stamos I (2012) A systematic approach for 2D-image to 3D-range registration in urban environments. Comput Vis Image Underst 116(1):25–37
28. Liu Z, Laganière R (2007) Context enhancement through infrared vision: a modified fusion scheme. Signal Image Video Process 1(4):293–301
29. Lowe DG (1991) Fitting parameterized three-dimensional models to images. IEEE Trans Pattern Anal Mach Intell 13(5):441–450
30. Lowe DG (1999) Object recognition from local scale-invariant features. In: Proceedings of the IEEE international conference on computer vision, pp 1150–1157
31. Lowe DG (2004) Distinctive image features from scale-invariant keypoints. Int J Comput Vis 60(2):91–110
32. Luhmann T, Ohm J, Piechel J, Roelfs T (2010) Geometric calibration of thermographic cameras. Int Arch Photogramm Remote Sens Spat Inf Sci XXXVIII-5:411–416
33. Ma Y, Soatto S, Košecká J, Sastry SS (2005) An invitation to 3-D vision: from images to geometric models. Springer, New York
34. Markov S, Birk A (2007) Detecting humans in 2D thermal images by generating 3D models. In: Hertzberg J, Beetz M, Englert R (eds) KI 2007: Advances in Artificial Intelligence. Lecture notes in artificial intelligence, vol 4667, Springer, Heidelberg, pp 293–307
35. Moreno-Noguer F, Lepetit V, Fua P (2007) Accurate non-iterative O(n) solution to the PnP problem. In: Proceedings of the IEEE international conference on computer vision, pp 1–8
36. Park C, Bae K-H, Choi S, Jung J-H (2008) Image fusion in infrared image and visual image using normalized mutual information. Proc SPIE 6968:69681Q-1–9
37. Parmehr EG, Fraser CS, Zhang C, Leach J (2013) Automatic registration of optical imagery with 3D lidar data using local combined mutual information. ISPRS Ann Photogramm Remote Sens Spat Inf Sci II-5/W2:229–234

38. Pelagotti A, Del Mastio A, Uccheddu F, Remondino F (2009) Automated multispectral texture mapping of 3D models. In: Proceedings of the European signal processing conference, pp 1215–1219
39. Pulli K, Abi-Rached H, Duchamp T, Shapiro LG, Stuetzle W (1998) Acquisition and visualization of colored 3D objects. In: Proceedings of the international conference on pattern recognition, pp 11–15
40. Prakash S, Pei YL, Caelli T (2006) 3D mapping of surface temperature using thermal stereo. In: Proceedings of the international conference on control, automation, robotics and vision, pp 1–4
41. Steger C (2001) Similarity measures for occlusion, clutter, and illumination invariant object recognition. In: Radig B, Florczyk S (eds) Pattern Recognition, DAGM2001. Lecture notes in computer science, vol 2191, Springer, Heidelberg, pp 148–154
42. Szeliski R (2011) Computer vision: algorithms and applications. Springer, London
43. Toth C, Ju H, Grejner-Brzezinska D (2011) Matching between different image domains. In: Stilla U, Rottensteiner F, Mayer H, Jutzi B, Butenuth M (eds) Photogrammetric image analysis, ISPRS Conference—Proceedings. Lecture notes in computer science, vol 6952, Springer, Heidelberg, pp 37–47
44. Ulrich M (2003) Hierarchical real-time recognition of compound objects in images. PhD thesis, Institut für Photogrammetrie und Kartographie, Technische Universität München, München, Germany
45. Vidas S, Moghadam P, Bosse M (2013) 3D thermal mapping of building interiors using an RGB-D and thermal camera. In: Proceedings of the IEEE international conference on robotics and automation, pp 2311–2318
46. Weinmann M (2013) Visual features—From early concepts to modern computer vision. In: Farinella GM, Battiato S, Cipolla R (eds) Advanced topics in computer vision. Advances in computer vision and pattern recognition, Springer, London, pp 1–34
47. Weinmann M, Jutzi B (2011) Fully automatic image-based registration of unorganized TLS data. Int Arch Photogramm Remote Sens Spat Inf Sci XXXVIII-5/W12:55–60
48. Weinmann M, Jutzi B (2012) A step towards dynamic scene analysis with active multi-view range imaging systems. Int Arch Photogramm Remote Sens Spat Inf Sci XXXIX-B3:433–438
49. Weinmann M, Hoegner L, Leitloff J, Stilla U, Hinz S, Jutzi B (2012) Fusing passive and active sensed images to gain infrared-textured 3D models. Int Arch Photogramm Remote Sens Spat Inf Sci XXXIX-B1:71–76
50. Weinmann M, Leitloff J, Hoegner L, Jutzi B, Stilla U, Hinz S (2014) Thermal 3D mapping for object detection in dynamic scenes. ISPRS Ann Photogramm Remote Sens Spat Inf Sci II-1:53–60
51. Wu Y, Hu Z (2006) PnP problem revisited. J Math Imaging Vis 24(1):131–141
52. Xue Z, Blum RS, Li Y (2002) Fusion of visual and IR images for concealed weapon detection. In: Proceedings of the international conference on image fusion, pp 1198–1205
53. Yao F, Sekmen A (2008) Multi-source airborne IR and optical image fusion and its application to target detection. In: Proceedings of the international symposium on advances in visual computing, pp 651–660
54. Zhang Z (2000) A flexible new technique for camera calibration. IEEE Trans Pattern Anal Mach Intell 22(11):1330–1334
55. Zitová B, Flusser J (2003) Image registration methods: a survey. Image Vis Comput 21(11): 977–1000

Chapter 6
3D Scene Analysis

Due to technological advancements in the last decade, terrestrial and mobile laser scanning systems are increasingly used for a fast, dense, and reliable 3D mapping. The acquired 3D representation of a scene in the form of 3D point cloud data as measured counterpart of physical object surfaces, in turn, may facilitate a variety of tasks among which particularly *3D scene analysis* in terms of automatically assigning 3D points a respective semantic class label has become a topic of great importance in photogrammetry, remote sensing, computer vision, and robotics. A respective assignment of semantic class labels, in turn, may represent an important prerequisite for further tasks such as consolidation of imperfect scan data, semantic perception for ground robotics, object detection and recognition, city modeling or urban accessibility analysis.

In this chapter, we consider a typical processing workflow for 3D scene analysis in detail, and we present a novel, fully automated, and versatile framework composed of four components addressing (i) neighborhood selection, (ii) feature extraction, (iii) feature selection, and (iv) classification. For each component, we consider a variety of approaches which allow applicability in terms of simplicity and reproducibility, so that end-users may easily apply the different components and do not require expert knowledge in the respective domains. Thereby, we also focus on performance in terms of accuracy and computational efficiency. Our main contributions address the issue of how to increase the distinctiveness of low-level geometric features and select the most relevant ones among these for 3D scene analysis. In an extensive evaluation involving 7 neighborhood definitions, 21 low-level geometric features, 7 approaches for feature selection, 10 classifiers of different categories and several benchmark datasets, we demonstrate that the selection of optimal neighborhoods for individual 3D points significantly improves the results of 3D scene analysis. Additionally, we show that the selection of adequate feature subsets may even further increase the quality of the derived results while significantly reducing both processing time and memory consumption.

Due to the consideration of optimal neighborhoods for individual 3D points, our framework is not tailored to a specific dataset, but principally designed to process 3D point clouds with a few millions of 3D points. Consequently, an extension toward

© Springer International Publishing Switzerland 2016
M. Weinmann, *Reconstruction and Analysis of 3D Scenes*,
DOI 10.1007/978-3-319-29246-5_6

data-intensive processing is desirable in order to also allow large-scale 3D scene analysis in case of huge 3D point clouds with possibly billions of points for a whole city. For this purpose, we present an extension which is based on an appropriate partitioning of the considered scene and thus allows a successive data processing in a reasonable amount of time without affecting the quality of the classification results. We demonstrate the performance of this extension on labeled benchmark datasets with respect to robustness, efficiency, and scalability.

Furthermore, we take into account that the different components in a typical processing workflow for 3D scene analysis have extensively, but separately been investigated in recent years, whereas the respective connection by sharing the results of crucial tasks across all components has not yet been addressed. This connection not only encapsulates the interrelated issues of neighborhood selection and feature extraction, but also the issue of how to involve spatial context in the classification step. Accordingly, we present a further extension which relies on involving individually optimized 3D neighborhoods for both the extraction of distinctive geometric features and the contextual classification of 3D point cloud data. In this regard, we demonstrate the beneficial impact of involving contextual information in the classification process and that using individual 3D neighborhoods of optimal size significantly increases the quality of the results for both individual and contextual classification.

Our framework has successfully undergone peer review and our main contributions address optimal neighborhood selection [139, 142] and feature relevance assessment [138, 139, 142, 144]. Furthermore, extensions of our framework have also successfully undergone peer review, and these extensions address large-scale capability [140, 141] as well as the integration of optimal neighborhood selection for contextual classification [143].

After providing a brief motivation for 3D scene analysis (Sect. 6.1), we reflect related work with respect to different research directions (Sect. 6.2). Subsequently, we explain the single components of our framework as well as respective methods in detail (Sect. 6.3). Since the framework is principally designed for 3D point clouds with a few millions of 3D points, we additionally present a respective extension of the framework toward large-scale 3D scene analysis (Sect. 6.4). Furthermore, we also consider an extension of our framework by involving contextual information (Sect. 6.5). In order to demonstrate the performance of our framework, we provide an extensive experimental evaluation on publicly available benchmark datasets (Sect. 6.6) and discuss the derived results (Sect. 6.7). Finally, we provide concluding remarks and suggestions for future work (Sect. 6.8).

6.1   Motivation and Contributions

Due to the recent technological advancements, modern scanning devices allow to efficiently acquire large amounts of geospatial data in the form of 3D point clouds and are therefore increasingly used for data acquisition. As a consequence of the increasing amount of captured data, a fully automated analysis of the captured 3D point clouds has become a topic of utmost importance in photogrammetry, remote sensing, computer vision, and robotics. Based on the acquired 3D point cloud data representing the measured counterpart of real-world object surfaces with a relatively high spatial resolution, recent investigations address a variety of different tasks such as 3D reconstruction [65, 88, 156], consolidation of imperfect scan data [20, 154], object detection [1, 16, 44, 61, 93, 104, 120, 129, 133, 150], extraction of roads and curbstones or road markings [10, 47, 157], urban accessibility analysis [119], recognition of power-line objects [60], extraction of building structures [132], large-scale city modeling [66, 102, 156], vegetation mapping [146], semantization of complex 3D scenes [9], or semantic perception for ground robotics [53]. However, many of these tasks rely on the results of an initial *3D point cloud classification* which aims at uniquely assigning a (semantic) class label (e.g., *ground*, *building* or *vegetation*) to each 3D point of a given 3D point cloud.

When addressing the task of 3D point cloud classification, we may take into account that many respective approaches reveal a similar structure of the processing workflow. Typically, 3D point cloud classification involves (i) the recovery of an appropriate local neighborhood for each 3D point, (ii) the extraction of suitable geometric features based on all 3D points within the respective local neighborhood, and (iii) the classification of all 3D points based on the respective features. In this regard, it has to be taken into account that—due to a lack of knowledge about the scene and/or data—often as many features as possible are extracted and used for classification. Among these features, however, some may be more relevant whereas others may be less suitable or even irrelevant. Accounting for this fact is of great importance since, in theory, many classifiers are considered to be insensitive to the given dimensionality, whereas redundant or irrelevant information has been proven to influence their performance in practice. Consequently, recent investigations also addressed the selection of meaningful features as an additional step between feature extraction and classification [22, 59, 79, 80]. Based on a generalized processing workflow consisting of all four steps, we may principally face a variety of challenges arising from the complexity of 3D scenes caused by irregular point sampling, varying point density and very different types of objects. Furthermore, it is to be expected that large 3D point clouds and many available features significantly increase the computational burden with respect to both processing time and memory consumption.

In this chapter, we intend to foster research in advanced 3D point cloud processing by considering the crucial task of 3D point cloud classification in detail. Focusing on a respective generalized processing workflow, we present a framework consisting of four components for neighborhood selection, feature extraction, feature selection, and classification (Fig. 6.1). Besides revisiting foundations and trends with respect to each component, we also provide new insights concerning the respective data processing. Whereas the different components in the processing workflow have extensively, but separately been investigated in recent years, the respective connection by sharing the results of crucial tasks across components has not yet been addressed. In this regard, we may take into account that neighborhood selection and feature extraction are strongly interrelated issues, since the distinctiveness of geometric features strongly depends on the respective neighborhood encapsulating those

Fig. 6.1 The proposed framework for 3D scene analysis: a 3D point cloud serves as input and the output consists of a (semantically) labeled 3D point cloud

3D points which are taken into consideration for feature extraction. We enrich our framework by providing a variety of approaches for different components and, based on an extensive evaluation involving three standard benchmark datasets, we are able to derive general statements on the suitability of the involved approaches.

In summary, our main contribution consists of a fully automated, efficient, generic, and general framework for 3D scene analysis which involves

- the recovery of individual neighborhoods of optimal size,
- the extraction of low-level geometric 3D and 2D features,
- different strategies for selecting relevant features,
- efficient approaches for a supervised classification based on the extracted features,
- an extension toward data-intensive processing, and
- an extension by involving contextual information

while preserving simplicity, efficiency, and reproducibility so that other researchers may easily extend our framework and end-users may easily apply the framework (or single components of the framework) without requiring expert knowledge in the different domains. Note that, although our main framework focuses on classifying each 3D point individually by only considering the respectively extracted geometric features, it also represents a basic requirement for either smooth labeling techniques or techniques involving contextual information within the local neighborhood of a considered 3D point, since both of them are based on the results of an individual classification.

Since only the spatial 3D geometry in terms of *XYZ*-coordinates serves as input, our framework is generally applicable for the semantic interpretation of 3D point cloud data obtained via different acquisition techniques such as terrestrial laser scanning (TLS), mobile laser scanning (MLS) or airborne laser scanning (ALS). Furthermore, our framework may also be applied for 3D point clouds captured with range cameras or 3D point clouds obtained via dense matching, i.e., via 3D reconstruction from images. Note that the selected feature subset may vary when considering datasets with different characteristics, while a significant impact of applying both optimal neighborhood size selection and feature selection may still be expected. We also want to point out that further extensions of our framework may be taken into account by involving additional features such as color/intensity or full-waveform features. Since such an extension will be independent from our scientific contribution, it is not in the scope of this work.

6.2 Related Work

Following the ideas presented in the previous section, we reflect related work on 3D scene analysis with respect to the fundamental components of neighborhood selection (Sect. 6.2.1), feature extraction (Sect. 6.2.2), feature selection (Sect. 6.2.3), and classification (Sect. 6.2.4).

6.2.1 Neighborhood Selection: Single-Scale Representation Versus Multi-Scale Representation

The first component of *neighborhood selection* assigns each 3D point $\mathbf{X}$ of a given 3D point cloud a respective local neighborhood which, in turn, is required in order to describe the local 3D structure around $\mathbf{X}$ via geometric features relying on the spatial arrangement of all 3D points within this local neighborhood. As a consequence, a respective neighborhood definition encapsulating all considered 3D points in a vicinity of $\mathbf{X}$ is required, for which different strategies may be applied. Among the strategies for defining the local neighborhood $\mathcal{N}$ around a given 3D point $\mathbf{X}$, the most commonly applied *neighborhood definitions* are represented by

- a spherical neighborhood definition $\mathcal{N}_s$ according to which the local neighborhood around a 3D point $\mathbf{X}$ is formed by all those 3D points within a sphere centered at $\mathbf{X}$ and parameterized with a fixed radius $r_s \in \mathbb{R}$ [71],
- a cylindrical neighborhood definition $\mathcal{N}_c$ according to which the neighborhood around a 3D point $\mathbf{X}$ is formed by all those 3D points whose 2D projections onto a plane (e.g., the ground plane) are within a circle centered at the projection of $\mathbf{X}$ and parameterized with a fixed radius $r_c \in \mathbb{R}$ [35], and
- a neighborhood definition $\mathcal{N}_k$ according to which the neighborhood around a 3D point $\mathbf{X}$ is formed by a fixed number of the $k \in \mathbb{N}$ closest neighbors of $\mathbf{X}$, where the distance metric relies on either 3D distances [73] or 2D distances [92].

Note that the third neighborhood definition $\mathcal{N}_k$ also results in a spherical neighborhood for $\mathbf{X}$ if 3D distances are evaluated in order to find the closest neighbors, but—in contrast to the first neighborhood definition $\mathcal{N}_s$—a variable radius and thus a variable absolute size is taken into account. In analogy to this, the third neighborhood definition $\mathcal{N}_k$ results in a cylindrical neighborhood for $\mathbf{X}$ if 2D distances are evaluated in order to find the closest neighbors and—in contrast to the second neighborhood definition $\mathcal{N}_c$—a variable radius and thus a variable absolute size is allowed.

Accordingly, all these commonly applied neighborhood definitions rely on the use of a single-scale parameter represented by either a radius $r \in \mathbb{R}$ or a number $k \in \mathbb{N}$ and, hence, a suitable parameterization has to be derived. In this regard, the straightforward solution consists of using a constant scale parameter, (i.e., either a fixed radius r or a constant value k) across all 3D points of a given 3D point cloud, whereby the scale parameter is typically selected based on heuristic or empiric knowledge about the scene and/or data and thus specific for each dataset. While the use of an identical scale parameter across all 3D points of a given 3D point cloud allows an efficient neighborhood selection, such a strategy does however not take into account that, intuitively, the neighborhood size and thus the scale parameter rather depend on local properties, (i.e., local 3D structure, point density, etc.) and, consequently, the scale parameter should not be defined as being identical across all considered 3D points. For an example, we may consider the characteristics of TLS or MLS data, where due to the process of data acquisition via line-of-sight instruments dense and accurate 3D point clouds with significant variations in point density may be expected.

In order to avoid the use of heuristic or empiric knowledge about the scene and/or data for neighborhood size selection, it is desirable to get rid of strong assumptions on local 3D neighborhoods, mainly represented by the use of an identical scale parameter across all 3D points of a given 3D point cloud. Accordingly, more recent investigations focus on the strategy of locally adapting the neighborhood size with respect to properties encapsulated in the local 3D structure and thus on introducing an optimal neighborhood size for each individual 3D point. As a result, neighborhood size selection is not only performed in a generic way, but also in a way which increases the distinctiveness of respectively derived geometric features. A consideration of respective approaches for *optimal neighborhood size selection* (where "optimality" refers to the respectively applied criterion) reveals that most of the presented approaches rely on the idea of a neighborhood based on the k closest 3D points and therefore focus on optimizing the scale parameter k for each individual 3D point. Thereby, the optimization may, for instance, be based on iterative schemes relating neighborhood size to curvature, point density, and noise of normal estimation [68, 82] which is particularly relevant for rather densely sampled and thus almost continuous surfaces. Accounting for more cluttered surface representations, other approaches rely on local surface variation [5, 99] or dimensionality-based scale selection [28]. While any of these approaches for deriving individual neighborhoods of optimal size causes additional effort in comparison to identically parameterized neighborhoods, the need for such concepts clearly becomes visible when considering the suitability of respective geometric features for neighborhoods of different size [4, 6].

While the consideration of *single-scale neighborhoods* with either identical or individual scale parameters across all 3D points of a given 3D point cloud has been addressed by different approaches, a further category of approaches focuses on deriving geometric features for multiple scales, since the strategy of varying the size of the local neighborhood and calculating all features for each scale considers additional information on how the local 3D geometry behaves across scales [17] which, in turn, may facilitate the discrimination between different classes and therefore improve the classification results [92]. In this regard, neighborhoods of small size allow to analyze fine details of the local 3D structure, whereas increasing the size of the local neighborhood is similar to applying a smoothing filter [99] as each individual 3D point will contribute less to the surface variation estimate if the neighborhood size is increased. When exploiting a scale-space representation in the form of *multi-scale neighborhoods* in order to extract geometric features over multiple scales, a training procedure may later be used to define which combination of scales allows the best separation of different classes [17]. However, even if the classification results may possibly be improved by applying such a scale-space representation, the number of involved scales and their distance in the scale-space are typically selected based on heuristic or empiric knowledge about the scene and/or data and, hence, specific for each dataset [17, 92, 117]. This raises the questions of how many scales should be involved, how suitable distances between neighboring scales may be derived and how their spacing should be defined (e.g., linear or nonlinear). Appropriately handling such issues by generic and fully automated techniques still remains challenging and has, to the best of our knowledge, not yet been addressed in literature. A solution could possibly be represented by the combination of generic approaches for selecting a single optimal scale with a generic multi-scale representation. For the latter, concepts from 2D scale-space theory could be applied in analogy to scale-space representations for detecting 3D interest points [152] or 3D interest regions [130]. Furthermore, it may be taken into account that the extraction of geometric features may not only rely on different scale parameters, but also on different entities such as points and regions [147–149] which may be motivated in analogy to the labeling paradigm, where the labels of single 3D points and their closest neighbors are exploited for a structured prediction.

Instead of relying on considerations on point-level, 3D scene analysis may also be based on the use of voxels or 3D segments. In this regard, it has for instance been proposed to exploit voxels carrying geometric, textural, and color information collected from airborne imagery and 3D point clouds derived via dense matching [42]. Furthermore, a multistage inference procedure exploiting a hierarchical segmentation based on voxels, blocks, and pillars has recently been presented [54]. While the use of voxels, blocks, and pillars relies on specific parameters to define the size of such entities, such parameters are typically selected based on heuristic or empiric knowledge about the scene and/or data. Hence, an alternative may consist of a 3D segmentation of the given 3D point cloud [134], where attributes of single 3D points may be exploited for a better separation of neighboring objects [95]. Additionally considering an automated scale selection, the combination of generic scale selection and voxelization has been proposed with an approach exploiting supervoxels derived via the oversegmentation of a 3D point cloud [72], where the size of the supervoxels is determined based on an iterative scheme involving curvature, point density, and noise and point intensities. In contrast to the use of voxels, blocks, and pillars, a 3D segmentation does not depend on an empirically or heuristically selected size of segments. However, a fully generic segmentation typically increases the computational burden significantly.

6.2.2   Feature Extraction: Sampled Features Versus Interpretable Features

The second component of *feature extraction* addresses the design of suitable features. Based on a respective definition and parameterization of local neighborhoods, a variety of approaches have been proposed in order to extract discriminative features from 3D point clouds with possibly varying point density. A classic approach for characterizing the local neighborhood around a given 3D point $\mathbf{X}$ exploits the strategy of deriving suitable 3D descriptors in the form of *spin images* [57] which are based on spinning a planar image patch around the surface normal at $\mathbf{X}$ (thus defining a local, cylindrical 3D neighborhood aligned with respect to the normal vector) and counting the number of points per pixel. A quite similar concept has been presented with the *3D shape context descriptor* [41] which, for a given 3D point $\mathbf{X}$, considers the distribution of other 3D points within a spherical neighborhood which, in turn, is oriented with respect to the normal vector at $\mathbf{X}$ and discretized into bins in order to form a histogram by counting the number of 3D points falling into each bin. In contrast, the *Signature of Histograms of OrienTations (SHOT) descriptor* [128] relies on exploiting a spherical grid centered at the 3D point $\mathbf{X}$, where each 3D bin of the grid is represented as weighted histogram of normals in order to achieve a better balance between descriptiveness and robustness. A powerful alternative to such sampling approaches is represented by *shape distributions* [94] which are based on randomly sampling geometric relations such as point distances, angles, areas, and volumes into histograms. While these shape distributions have originally been introduced for describing specific geometric properties of a complete 3D model, a respective adaptation for characterizing the local neighborhood of a 3D point $\mathbf{X}$ has recently been presented [6]. Furthermore, it has been proposed to exploit *Point Feature Histograms (PFHs)* [109, 110] which focus on sampling geometric relations such as point distances and angular variations between the closest neighbors of each considered 3D point $\mathbf{X}$ relative to the local normal vector at $\mathbf{X}$ into histograms. However, since all these approaches rely on concatenating histogram entries to a feature vector, single entries of the resulting feature vector are hardly interpretable.

Consequently, a variety of approaches for 3D scene analysis relies on interpretable features. In this regard, the most commonly applied approach is based on the idea of exploiting the *3D structure tensor* which represents the 3D covariance matrix derived from the 3D coordinates of all 3D points within the local neighborhood of a considered 3D point $\mathbf{X}$. The 3D structure tensor thus encapsulates information about the local 3D structure at $\mathbf{X}$ and this information may, for instance, be described by considering the three eigenvalues of the 3D structure tensor. More specifically, an analytical consideration of these eigenvalues has been proposed in order to characterize specific shape primitives [58]. Furthermore, the three eigenvalues of the 3D structure tensor may be exploited in order to derive a set of local 3D shape features [99, 145] which addresses more intuitive descriptions (e.g., about linear, planar, or volumetric structures), and is therefore commonly applied in LiDAR data processing. While such a feature set offers an intuitive description of local 3D structures, it is typically extended with other geometric features which may be based on angular statistics [85, 86], height and local plane characteristics [80], height characteristics and curvature properties [115, 116], descriptors involving surface properties, slope, height characteristics, vertical profiles and 2D projections [49] or point distances and height differences [136]. Moreover, eigenvalue-based 3D features are often combined with additional full-waveform and echo-based features [22, 80, 91, 117, 136].

Note that, in contrast to a consideration of the 3D covariance matrix encoding information about the spatial arrangement of 3D points within a local neighborhood, the use of covariance matrices of higher dimension has been proposed in order to combine features such as angular measures and point distances to a compact representation [34]. In this context, it has also been pointed out that such a representation may easily be extended by taking into account further features, e.g., arising from radiometric information. However, the covariance matrices do not adhere to the Euclidean geometry, but span a Riemannian space with an associated Riemannian distance metric [34], e.g., the Geodesic distance [37] or the log-Euclidean Riemannian metric as used in [24, 34].

6.2.3 Feature Selection: All Features Versus Relevant Features

The third component of *feature selection* addresses the fact that, when addressing 3D scene analysis, typically as many features as possible are exploited in order to compensate a lack of knowledge about the scene and/or data. However, it has to be taken into account that some of these features may be more and others less relevant, and some features even might be redundant. Accounting for this fact is of great importance since, in theory, many classifiers are considered to be insensitive to the given dimensionality, whereas redundant or irrelevant information has been proven to influence their performance in practice. The reason behind such observations becomes obvious when trying to learn the "nature" of considered data from a finite number of training examples given in a high-dimensional feature space. Typically, a huge number of training examples are required in order to ensure that all possibly occurring effects are representatively covered. For a limited number of training examples, however, an increasing amount of features will first increase the predictive power of a classifier, reach a maximum predictive power of this classifier for a specific number of features (which depends on the number of training examples and the type of the involved classifier) and finally decrease the predictive power of the classifier again. This decrease is commonly referred to as the *Hughes phenomenon* [55] and it mainly results from an inaccurate estimation of class statistics [7]. Respective observations revealing the Hughes phenomenon may also be found in the context of 3D point cloud classification [79, 80] or in the context of a classification of hyperspectral remote sensing images [81].

Since a large number of involved features significantly increase the computational burden and thereby may even reduce the predictive power of a classifier, more and more attention has been paid to feature selection techniques for finding compact and robust subsets of relevant and informative features in order to gain predictive accuracy, improve computational efficiency with respect to both time and memory consumption, and retain meaningful features [50, 51, 74, 111]. While a variety of respective approaches has been presented in the literature, they all follow one of the three fundamental strategies for feature selection and may therefore be categorized into *filter-based methods*, *wrapper-based methods*, and *embedded methods*. The main characteristics, advantages, and disadvantages of these different strategies are provided in Table 6.1. For more details, we refer to a comprehensive survey on feature selection techniques [111].

While feature selection techniques have typically been introduced for high-dimensional classification problems, where datasets with tens or hundreds of thousands of variables are available as, e.g., for text processing of internet documents, bioinformatics and combinatorial chemistry [50, 111], respective techniques have

Table 6.1 Categorization of feature selection (FS) techniques according to their main characteristics [111, 142]

Strategy	Advantages	Disadvantages	Examples
Filter-based FS (univariate)	Simple	No feature dependencies	Fisher score
	Fast	No interaction with the classifier	Information gain
	Classifier-independent		Symmetrical uncertainty
Filter-based FS (multivariate)	Classifier-independent	Slower than univariate techniques	ReliefF
	Models feature dependencies	No interaction with the classifier	CFS
	Faster than wrapper-based FS		FCBF
			mRMR
Wrapper-based FS	Interaction with the classifier	Classifier-dependent	SFS
	Models feature dependencies	Computationally intensive	SBE
			Simulated annealing
Embedded FS	Interaction with the classifier	Classifier-dependent	Random forests
	Models feature dependencies		AdaBoost
	Faster than wrapper-based FS		

only rarely been applied in the context of 3D point cloud processing. Aiming at the use of such techniques, we have to take into account different aspects. As both wrapper-based methods and embedded methods involve a classifier, they generally tend to yield a better performance than filter-based methods which focus on classifier-independent schemes. More specifically, wrapper-based methods typically involve a classifier either via *Sequential Forward Selection* (*SFS*) which is based on finding the feature yielding the highest predictive accuracy and successively adding the feature that improves performance the most, or via *Sequential Backward Elimination* (*SBE*) which is based on starting with the complete feature set and repeatedly deleting the feature that reduces performance the least [59, 79, 80]. In this regard, the results for SFS and SBE are likely to differ, since SFS tends to yield smaller feature subsets, whereas SBE tends to provide larger feature subsets. However, such wrapper-based methods cause a rather high computational effort due to the involved classifier which has to be trained independently for all considered feature subsets. In contrast to wrapper-based methods, embedded methods (e.g., Random Forests [15] or AdaBoost [38]) are able to deal with exhaustive feature sets as input and let the classifier internally select a suitable feature subset during the training phase [22, 48, 127], which is much more efficient than training a classifier several times. As a consequence of relying on a classifier, however, both wrapper-based methods and embedded methods provide feature subsets which are optimized with respect to the involved classifier (and thus also with respect to its type and its settings), and results may significantly vary when using different classifiers. In order to overcome feature subsets which are only optimized with respect to a specific classifier, filter-based methods rely on classifier-independent schemes for deriving versatile feature subsets and therefore tend to provide a slightly weaker performance. More specifically, such filter-based methods only exploit a score function which is evaluated based on given training data which, in turn, results in simplicity and efficiency. Since the score function may be based on (i) using only feature-class relations in order to determine relevant features or (ii) using both feature-class relations and feature-feature relations in order to determine relevant features and thereby remove redundancy, such filter-based methods may be categorized into *univariate* and *multivariate* techniques.

6.2.4 Classification: Individual Feature Vectors Versus Contextual Information

The fourth component of *classification* aims at uniquely assigning a (semantic) class label to each 3D point of a given 3D point cloud. Even though the three already explained components are important to obtain adequate results, most of the effort with respect to 3D scene analysis has addressed the design and improvement of the classification procedure. For being able to reason about the suitability of an involved classifier, the strategy of supervised classification is typically applied which relies on exploiting a set of training examples in order to train the classifier in a way that it may afterward generalize the learned information to new, unseen data. Respective approaches for supervised classification may generally be categorized with respect to (i) an *individual classification* where each 3D point is classified by only exploiting the respective feature vector or (ii) a *contextual classification* where each 3D point is classified by exploiting the respective feature vector as well as contextual information derived from the feature vectors and class labels of neighboring 3D points.

For 3D scene analysis, the strategy of an individual classification of 3D points based on the respective feature vectors has been followed by applying standard classification approaches such as classical maximum likelihood classifiers based on Gaussian mixture models [68, 69], Support Vector Machines [75, 79, 80, 118], AdaBoost [76], Random Forests [22], a cascade of binary classifiers [19], and Bayesian discriminant classifiers [59]. Such an individual classification may be carried out very efficiently, and respective approaches are publicly available in a variety of software tools and thus easily applicable. When focusing on an individual classification of 3D points, however, we may expect that the derived result, (i.e., the derived labeling) reveals a noisy behavior since this strategy does not take into account that the (semantic) class labels of neighboring 3D points tend to be correlated.

In order to account for probably correlated class labels of nearby 3D points, a contextual classification not only relies on the respective feature vectors but also on relationships among 3D points within a local neighborhood, whereby the relationships have to be inferred from the training data by modeling interactions between neighboring 3D points. In this regard, it is important to notice that the local neighborhood for inferring relationships among 3D points is typically different from the one used for feature extraction. In the scope of 3D scene analysis, the strategy of a contextual classification of 3D points has been followed by applying Associative Markov Networks [2, 85–87] and non-Associative Markov Networks [89, 123, 124], Conditional Random Fields [72, 91, 92, 115, 117], Simplified Markov Random Fields [78], multistage inference procedures focusing on point cloud statistics and relational information over different scales [147], and spatial inference machines modeling mid- and long-range dependencies inherent in the data [125]. While some methods such as the one presented in [123] are based on the consideration of 3D point cloud segments, other methods such as the one presented in [92] directly classify 3D points. Since segment-based methods strongly depend on the quality of the results of the applied algorithm for 3D point cloud segmentation, point-based techniques seem to be favorable. Generally, however, approaches based on a contextual classification tend to provide better classification results than approaches based on an individual classification, since they focus on a rather smooth labeling and the derived results do therefore not reveal a noisy behavior.

While a spatially smooth labeling is desirable, applying the strategy of a contextual classification of 3D points typically increases the computational effort drastically, which is due to the modeling of interactions between neighboring 3D points, and therefore often tends to be impracticable. Furthermore, it should be taken into account that an exact inference is computationally intractable, since the given training data is limited and the modeling of relationships across the whole training data results in too many degrees of freedom. In order to overcome such issues, it has been proposed to consider approximate inference techniques instead of exact inference techniques. In this regard, we may have a closer look on the modeling of interactions between neighboring 3D points, where a modeling of long-range interactions is feasible in order to improve the predictive power of the classifier and rather time-consuming, while only a modeling of short-range interactions is also feasible in order to improve computational efficiency, but maybe not as much. Accordingly, approximate inference techniques may for instance rely on only inferring spatial relationships among 3D points within a local 3D neighborhood and thus rather short-range dependencies inherent in the data. Note that a respective approach may easily be motivated by the fact that the correlation of class labels is likely to be encapsulated in a local 3D neighborhood. In this regard, it has been shown that even such techniques quickly reach their limitations if the considered local neighborhood becomes too large. However, there is no general indication toward an optimal inference strategy yet which allows recommendations concerning the choice of an optimal context model or the modeling of short-range, mid-range and long-range interactions. For approaches only relying on a modeling of short-range interactions, an efficient alternative is represented by a straightforward labeling without smoothness constraint (i.e., via individual classification) and subsequently enforcing the desirable smooth labeling of nearby 3D points via probabilistic relaxation [83] or smooth labeling techniques [114].

In the scope of this chapter, we take into account that many contextual classification approaches and all smoothing techniques rely on the results of an individual classification of all 3D points of a given 3D point cloud. Consequently, it is desirable to further investigate sources for potential improvements in individual classification and to reason about their impact on a subsequent exploitation of either contextual information or local smoothing.

6.3 A Novel Framework for 3D Scene Analysis

For 3D scene analysis in terms of uniquely assigning a (semantic) class label to each 3D point of a given 3D point cloud, we propose a fully automated and generic framework which is based on a generalized processing workflow as shown in Fig. 6.1. In the following subsections, we explain the involved components addressing neighborhood selection (Sect. 6.3.1), feature extraction (Sect. 6.3.2), feature selection (Sect. 6.3.3), and classification (Sect. 6.3.4) as well as respective methods in detail.

Our main scientific contribution is based on characterizing each individual 3D point of a given 3D point cloud with an individual neighborhood of optimal size. This allows an extraction of highly distinctive geometric 3D and 2D features which is pursued subsequently. Based on the derived features, we account for the Hughes phenomenon [55] and therefore aim at quantifying the relevance of features in order to select compact subsets of versatile features which also reduces the computational burden with respect to processing time and memory consumption which, in turn, represents a particularly important prerequisite for large-scale 3D scene analysis.

Finally, training examples and a suitable feature (sub)set are provided as input for a supervised classification scheme.

6.3.1   Neighborhood Selection

As explained in Sect. 6.2.1, the component of *neighborhood selection* may focus on either single-scale approaches or multi-scale approaches. On the one hand, involving a multi-scale approach seems to be desirable in order to also account for the behavior of features across different scales by providing all feature values to a classifier and thus delaying the decision about the suitability of a specific neighborhood to this classifier. On the other hand, however, a multi-scale approach requires decisions, e.g., on how the scale-space is designed, how many scales are involved and how the distance between scales is determined. As such important decisions typically rely on heuristic or empiric knowledge about the scene and/or data [17, 92, 117], the respectively derived settings are specific for each dataset. In contrast, a single-scale approach may rely on a generic scale selection which even allows a generalization between datasets, since the neighborhood size is adapted to the local 3D structure and variations in point density may be handled as well. For these reasons, we focus on a single-scale approach in the scope of our work.

Without loss of generality, we consider 3D points $\mathbf{X} = (X, Y, Z)^T \in \mathbb{R}^3$ of a given 3D point cloud, where the XY-plane represents a horizontally oriented plane and, consequently, the Z-axis is oriented along the vertical direction. Note that respective data is typically provided in the commonly available MLS point cloud datasets and in georeferenced data. The goal of neighborhood selection consists of recovering the neighboring 3D points for $\mathbf{X}$ which, in turn, involves (i) a suitable neighborhood definition, (ii) an efficient recovery of neighboring 3D points, and (iii) an adequate parameterization in terms of neighborhood size. These aspects are addressed in the following subsections.

6.3.1.1   Neighborhood Definition

When focusing on a single-scale approach, the selection of a suitable neighborhood definition (Fig. 6.2) and its respective scale parameter certainly depends on the characteristics of the given 3D point cloud data. While applying the spherical neighborhood definition $\mathcal{N}_s$ or the cylindrical neighborhood definition $\mathcal{N}_c$ typically involves empiric or heuristic knowledge about the scene and/or data in order to determine a suitable radius, the neighborhood definition $\mathcal{N}_k$ based on the k closest neighbors of a given 3D point $\mathbf{X}$ offers more flexibility with respect to the absolute neighborhood size and is more adaptive to varying point density. With the intention of providing a generally versatile framework, we explicitly want to avoid including prior knowledge about the scene and/or data (which would restrict our framework to a specific dataset) and therefore involve the neighborhood definition $\mathcal{N}_k$. Thereby, due to the

 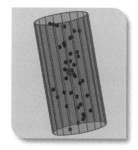 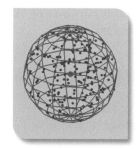

Fig. 6.2 Visualization of different neighborhood definitions for a given 3D point **X** (indicated by a *green dot*) according to [46, 58]: *spherical* neighborhood definition (*left*), *cylindrical* neighborhood definition (*center*) and neighborhood definition formed by the k closest neighbors when considering the Euclidean 3D distance (*right*). All other 3D points within the respective neighborhood are indicated by *red dots*

consideration of densely sampled TLS/MLS point cloud data, we evaluate the closest neighbors based on 3D distances and thus obtain spherical neighborhoods with flexible radius. Consequently, we need to involve a strategy for an efficient recovery of the closest neighbors of **X** as well as a strategy for a generic selection of a suitable number $k \in \mathbb{N}$ of considered neighbors.

6.3.1.2 Neighborhood Recovery

As a consequence of the selected neighborhood definition, a computationally intensive part consists of finding the closest neighbors for each 3D point **X** of a given 3D point cloud. This *nearest neighbor search* may formally be described as follows: given a point set $\mathscr{S} = \{\mathbf{X}_1, \ldots, \mathbf{X}_{N_P}\}$ consisting of $N_P \in \mathbb{N}$ points $\mathbf{X}_i$ in a 3D Euclidean vector space, those points $\mathbf{X}_j \in \mathscr{S}$ should efficiently be recovered which are closest to a given query point $\mathbf{X} \in \mathbb{R}^3$. In order to find these closest neighbors for **X**, the commonly used approach is based on a KD-tree [39]. Generally, a KD-tree represents a compact hierarchical data structure for point sets sampled from a K-dimensional manifold and thus allows an efficient recovery of neighboring points. A point in a KD-tree with N points may thus be localized with an average complexity of $O(\log N)$ and, in the worst case, with a complexity of $O(N)$ [135]. If required, a further increase in computational efficiency may be achieved by replacing the exact nearest neighbor search by an approximate nearest neighbor search [3] which, in turn, introduces little loss in accuracy since nonoptimal neighbors may be returned. For more details on respective approaches, we refer to an extensive and valuable survey on data structures [112]. A good trade-off between computational efficiency and accuracy has been presented with the Fast Library for Approximate Nearest Neighbors (FLANN) [84] which is based on either (i) searching hierarchical K-means trees with a priority search order or (ii) using multiple randomized KD-trees. Since the FLANN is publicly available (e.g., in OpenCV) and may thus also easily be applied in other software tools (e.g., Matlab), we apply it for nearest neighbor search in our framework.

6.3.1.3 Neighborhood Parameterization

Besides an efficient nearest neighbor search, it is desirable to automatically select an optimal parameterization for the selected neighborhood definition which, in our case, is based on the k closest neighbors. Consequently, we need to find a suitable value for the scale parameter k (which is also commonly referred to as *scale*), where the simplest, straightforward, and hence commonly used approach consists of selecting a fixed value which is identical for all 3D points of the considered 3D point cloud. However, as outlined in Sect. 6.2.1, we may intuitively rather prefer a choice where the scale parameter k is more flexible and thus allowed to vary within a 3D point cloud, since k certainly depends on the local properties of the considered 3D point cloud which are represented by the respective 3D structure and the local point density. Consequently, we focus on a generic selection of individual neighborhoods described by an optimized scale parameter k for each 3D point $\mathbf{X}$ which, in turn, completely avoids the use of prior knowledge about the scene and/or data.

In order to locally adapt the neighborhood size of each 3D point $\mathbf{X}$ to the given 3D point cloud data, where the optimization relies on a specific energy function E, we follow seminal work addressing the selection of an optimal value for the scale parameter k based on fundamental geometric properties of the given 3D point cloud [28, 99]. Respective approaches typically exploit energy functions relying on the well-known *3D structure tensor* $\mathbf{S}_{3D} \in \mathbb{R}^{3\times3}$ with

$$\mathbf{S}_{3D} = \frac{1}{k+1} \sum_{i=0}^{k} \left(\mathbf{X}_i - \bar{\mathbf{X}}\right) \left(\mathbf{X}_i - \bar{\mathbf{X}}\right)^T \tag{6.1}$$

which simply represents the 3D covariance matrix derived from the 3D coordinates of a given 3D point $\mathbf{X} = \mathbf{X}_0$ and its k closest neighbors $\mathbf{X}_i$ with $i = 1, \dots, k$. In this equation, the geometric center $\bar{\mathbf{X}}$ is defined as

$$\bar{\mathbf{X}} = \frac{1}{k+1} \sum_{i=0}^{k} \mathbf{X}_i \tag{6.2}$$

and it may hence slightly vary from the considered 3D point $\mathbf{X}$. Since the 3D structure tensor represents a symmetric positive semidefinite matrix, its three eigenvalues exist, are nonnegative, and correspond to an orthogonal system of eigenvectors. Furthermore, we may assume that there may not necessarily be a preferred variation with respect to the eigenvectors so that we have to consider the general case of a structure tensor with rank 3. As a consequence, the three eigenvalues λ_1, λ_2, and λ_3 with $\lambda_1, \lambda_2, \lambda_3 \in \mathbb{R}$ and $\lambda_1 \geq \lambda_2 \geq \lambda_3 \geq 0$ indicate the extent of a 3D ellipsoid along its three principal axes. Thus, the eigenvalues encapsulate valuable information for characterizing the local 3D shape at a considered 3D point **X** and they may therefore be exploited in order to deduce a suitable description of the local 3D shape. In the context of optimal neighborhood size selection, two approaches have been presented which exploit the eigenvalues of the 3D structure tensor in order to define a respective energy function E.

The first approach exploits the eigenvalues of the 3D structure tensor in order to estimate the *local surface variation* [5, 99] according to

$$C_\lambda = \frac{\lambda_3}{\lambda_1 + \lambda_2 + \lambda_3} \tag{6.3}$$

which is also referred to as *change of curvature* [108]. By starting with small values of the scale parameter $k \in \mathbb{N}$ and successively increasing the neighborhood size by adding the next-closest neighbor to the local neighborhood, a critical neighborhood size and thus a respective value for k corresponds to a significant increase of C_λ, since occurring jumps indicate strong deviations in the normal direction [5, 99].

The second approach is referred to as *dimensionality-based scale selection* [28] as it exploits the eigenvalues of the 3D structure tensor in order to derive the *dimensionality features* represented by *linearity L_λ*, *planarity P_λ*, and *scattering S_λ* according to

$$L_\lambda = \frac{\lambda_1 - \lambda_2}{\lambda_1} \tag{6.4}$$

$$P_\lambda = \frac{\lambda_2 - \lambda_3}{\lambda_1} \tag{6.5}$$

$$S_\lambda = \frac{\lambda_3}{\lambda_1} \tag{6.6}$$

and these dimensionality features represent 1D, 2D, and 3D features. In these three equations, the normalization by the largest eigenvalue λ_1 results in dimensionality features $L_\lambda, P_\lambda, S_\lambda \in \mathbb{R}$ with $L_\lambda, P_\lambda, S_\lambda \in [0, 1]$ which furthermore sum up to 1. Thus, the dimensionality features satisfy two of the three probability axioms according to [62]. Taking into account that quasiprobability distributions generally relax the third axiom addressing the junction of mutually disjoint random events, we may consider the dimensionality features as the "probabilities" of a 3D point to be labeled as 1D, 2D, or 3D structure [28]. Accordingly, the task of finding a suitable neighborhood size may be transferred to favoring one dimensionality the most (thus favoring dimensionality features which are as dissimilar as possible from each other) which, in turn, corresponds to minimizing an energy function E_{dim} expressing a measure of unpredictability via the Shannon entropy [122] according to

$$E_{dim} = -L_\lambda \ln(L_\lambda) - P_\lambda \ln(P_\lambda) - S_\lambda \ln(S_\lambda) \tag{6.7}$$

across different scales $k \in \mathbb{N}$:

$$k_{\text{opt,dim}} = \arg\min_{k} E_{\text{dim}}(k) \qquad (6.8)$$

Note that, according to the original implementation [28], the neighborhood radius r has been taken into account and the interval $r \in [r_{\min}, r_{\max}]$ has been sampled in 16 scales, whereby the radii are not linearly increased since the radius of interest is usually closer to $r_{\min}$. Consequently, the optimal neighborhood size corresponds to the radius which yields the minimum Shannon entropy. However, since the values $r_{\min}$ and $r_{\max}$ have been selected based on various characteristics of the given 3D point cloud data, they are specific for each dataset. In order to avoid such a heuristic or empiric parameter selection, we propose to slightly adapt the original implementation by directly varying the scale parameter k instead of considering specified radii. In order to minimize the defined energy function E_{dim} and thus obtain an optimal scale parameter $k_{\text{opt,dim}}$, we consider all integer values in the interval $[k_{\min}, k_{\max}]$. Thereby, we consider relevant statistics to start with $k_{\min} = 10$ neighbors—which is in accordance with the original approach [28]—and we successively increase the scale parameter with a step size of $\Delta k = 1$ up to a relatively high number of $k_{\max} = 100$. Compared to the original implementation with 16 scales, our procedure results in an increase of the computational effort with respect to processing time and memory consumption, but remains generic since we may increase the upper boundary arbitrarily if this should be necessary (which we will have to verify by carrying out respective experiments).

Inspired by dimensionality-based scale selection, where optimal neighborhood size selection is based on the idea that the optimal neighborhood size favors one dimensionality the most, we propose a third approach which directly exploits the eigenvalues of the 3D structure tensor in order to estimate the order/disorder of 3D points within the local 3D neighborhood. For this purpose, we take into account the definition of quasiprobability distributions and normalize the three eigenvalues λ_1, λ_2, and λ_3 by their sum Σ_λ in order to obtain the normalized eigenvalues $e_i = \lambda_i / \Sigma_\lambda$ with $i \in \{1, 2, 3\}$. Since the normalized eigenvalues $e_i \in [0, 1]$ sum up to 1, we may consider them as "probabilities" and, consequently, define an energy function E_λ expressing the measure of *eigenentropy* via the Shannon entropy according to

$$E_\lambda = -e_1 \ln(e_1) - e_2 \ln(e_2) - e_3 \ln(e_3) \qquad (6.9)$$

where the occurrence of eigenvalues identical to zero has to be avoided by adding an infinitesimal small value ε. The eigenentropy E_λ represents a measure describing the order/disorder of 3D points within the local 3D neighborhood, and our approach which we refer to as *eigenentropy-based scale selection* is based on the idea of favoring a minimum disorder of 3D points which, in turn, corresponds to minimizing the energy function E_λ across different scales $k \in \mathbb{N}$:

$$k_{\text{opt},\lambda} = \arg\min_{k} E_\lambda(k) \qquad (6.10)$$

In order to minimize the defined energy function E_λ and thus obtain an optimal scale parameter $k_{\mathrm{opt},\lambda}$, we follow the idea of our adaptation introduced for dimensionality-based scale selection and hence vary the scale parameter k between $k_{\min} = 10$ and $k_{\max} = 100$ with $\Delta k = 1$. Note that the optimal neighborhood size is thus again determined by varying the scale parameter k and selecting the value which yields the minimum Shannon entropy.

Both dimensionality-based scale selection and eigenentropy-based scale selection are generally applicable in order to select individual neighborhoods of optimal size, since they neither involve parameters which are specific for a dataset nor rely on the assumption of particular shapes being present in the observed scene. Our contribution thus focuses on feature design in terms of deriving distinctive geometric features from individual neighborhoods of optimal size and, consequently, our main claim (which we will have to verify by carrying out respective experiments) is that the distinctiveness of geometric features calculated for a considered 3D point **X** from its neighboring 3D points is increased when involving individually optimized neighborhoods. In this regard, we may also have to take into account the additional computational effort caused by optimal neighborhood size selection and furthermore conclude if the consideration of individual neighborhoods of optimal size justifies the additional effort.

6.3.2 Feature Extraction

Once a neighborhood size has been specified for each 3D point **X** of a given 3D point cloud, the next step consists of deriving respective features in order to obtain a suitable description in the form of feature vectors. Since still many datasets only contain geometric information in the form of spatial 3D coordinates, we focus on describing each 3D point **X** only via geometric features which are implicitly preserved in the spatial arrangement of all 3D points within its local neighborhood. Thus, very different types of 3D point clouds may serve as input for our framework, e.g., 3D point clouds acquired via terrestrial/mobile laser scanning or 3D point clouds generated via dense matching. Especially these types of 3D point clouds provide an adequate point density for an appropriate representation of object surfaces as measured or generated counterpart of the real world. The derived feature vectors may easily be extended by considering further information in the form of intensity/color or full-waveform features, but this is not in the scope of our work since it does not influence the presented scientific contribution.

In order to extract respective geometric features for describing a given 3D point **X** based on its local neighborhood, we take into account that neighborhood selection and feature extraction are interrelated issues, since the distinctiveness of geometric features strongly depends on the respective neighborhood encapsulating those 3D points which are taken into consideration for feature extraction. In this regard, some features will probably be more distinctive for larger neighborhoods (e.g., features addressing planarity or curvature), whereas other features will probably be more distinctive for smaller neighborhoods (e.g., features addressing fine details of the local 3D structure). While principally a respective neighborhood size may be selected for each individual feature, such an approach tends to be intractable since a desired generic neighborhood selection with respect to different cues drastically increases the computational burden. As a consequence, we want to avoid optimizing the neighborhood size for each single feature and therefore rely on the concept of either dimensionality-based scale selection or eigenentropy-based scale selection while assuming that the consideration of locally adaptive neighborhoods is sufficient in order to retrieve distinctive features. Thus, we focus on the concept of decoupling neighborhood selection and feature extraction and treat feature extraction independent from neighborhood selection since an interaction of these components in terms of optimization is not taken into account.

Consequently, based on the respective local 3D neighborhood derived via either dimensionality-based scale selection or eigenentropy-based scale selection, we derive basic geometric 3D properties as well as local 3D shape features for each 3D point $\mathbf{X}$ of the considered 3D point cloud (Sect. 6.3.2.1). Furthermore, we also involve 2D neighborhoods based on a 2D projection and thus derive basic geometric 2D properties, local 2D shape features and features based on a discrete 2D accumulation map (Sect. 6.3.2.2). In total, this yields a feature set consisting of 21 low-level geometric 3D and 2D features which are categorized in Table 6.2 and briefly described in the following subsections. Introducing the definition of a *site*, which is represented by a 3D point in our case, we concatenate the respective 21 extracted features to a vector and thus obtain the site-wise feature vector $\mathbf{f}$ describing the 3D point $\mathbf{X}$. Since the derived geometric features address different quantities with possibly different units, we involve a normalization $[\cdot]_{\text{norm}}$ across all sites and thus all feature vectors, whereby the values of each dimension are respectively mapped to the interval $[0, 1]$.

Table 6.2 Categorization of the 21 involved features

Geometric 3D properties	Local 3D shape features	Geometric 2D properties	Local 2D shape features	Features based on a 2D accumulation map
H	L_λ	$r_{k\text{-NN},2D}$	$\Sigma_{\lambda,2D}$	M
$r_{k\text{-NN},3D}$	P_λ	D_{2D}	$R_{\lambda,2D}$	ΔH
D_{3D}	S_λ			σ_H
V	O_λ			
$\Delta H_{k\text{-NN},3D}$	A_λ			
$\sigma_{H,k\text{-NN},3D}$	E_λ			
	Σ_λ			
	C_λ			

As a consequence, we may characterize each 3D point $\mathbf{X}$ by a 21-dimensional feature vector $\mathbf{f} \in \mathbb{R}^{21}$:

$$
\begin{aligned}
\mathbf{f} = \big[& H, r_{k\text{-NN},3D}, D_{3D}, V, \Delta H_{k\text{-NN},3D}, \sigma_{H,k\text{-NN},3D}, \\
& L_\lambda, P_\lambda, S_\lambda, O_\lambda, A_\lambda, E_\lambda, \Sigma_\lambda, C_\lambda, \\
& r_{k\text{-NN},2D}, D_{2D}, \Sigma_{\lambda,2D}, R_{\lambda,2D}, M, \Delta H, \sigma_H \big]^T_{\text{norm}}
\end{aligned}
\tag{6.11}
$$

6.3.2.1 3D Features

A variety of *3D features* may directly be derived by considering basic geometric 3D properties of the local neighborhood and local 3D shape features arising from the spatial arrangement of 3D points within the neighborhood.

Geometric 3D Properties With the considered notation of a 3D point $\mathbf{X} = (X, Y, Z)^T \in \mathbb{R}^3$ of a given 3D point cloud, we have specified that the XY-plane represents a horizontally oriented plane and that the Z-axis is oriented along the vertical direction. Thus, in the context of 3D scene analysis, valuable information about a given 3D point $\mathbf{X}$ might be represented by its height value $H = Z$. Further basic geometric 3D properties of the considered local neighborhood are given by the radius $r_{k\text{-NN},3D}$ of the spherical neighborhood encapsulating the k closest neighbors, the *local point density D_{3D}* given by

$$
D_{3D} = \frac{k + 1}{\frac{4}{3} \pi \, r^3_{k\text{-NN},3D}}
\tag{6.12}
$$

and the *verticality* $V = 1 - n_Z$ [29] derived from the vertical component n_Z of the normal vector $\mathbf{n} \in \mathbb{R}^3$. Note that the normal vector is represented by the eigenvector corresponding to the smallest eigenvalue of the respective 3D structure tensor. In addition, the maximum difference $\Delta H_{k\text{-NN},3D}$ and standard deviation $\sigma_{H,k\text{-NN},3D}$ of height values within the neighborhood may be considered as interesting properties of the considered local neighborhood.

Local 3D Shape Features In contrast to considering basic geometric 3D properties of the local neighborhood, we may also exploit all 3D points within this neighborhood in order to derive the 3D structure tensor $\mathbf{S}_{3D}$ and its normalized eigenvalues e_i with $i \in \{1, 2, 3\}$ which encode information about the spatial distribution of 3D points within the local neighborhood. Based on these normalized eigenvalues, we may define a set of eight *local 3D shape features* [99, 145] which is composed of the dimensionality features of *linearity L_λ*, *planarity P_λ*, and *scattering S_λ* (Figs. 6.3 and 6.4) as well as further features represented by *omnivariance O_λ*, *anisotropy A_λ*, *eigenentropy E_λ*, sum Σ_λ of eigenvalues, and *change of curvature C_λ*. Hence, for each 3D point $\mathbf{X}$ and its respective k closest neighbors, these features are directly derived from the normalized eigenvalues according to

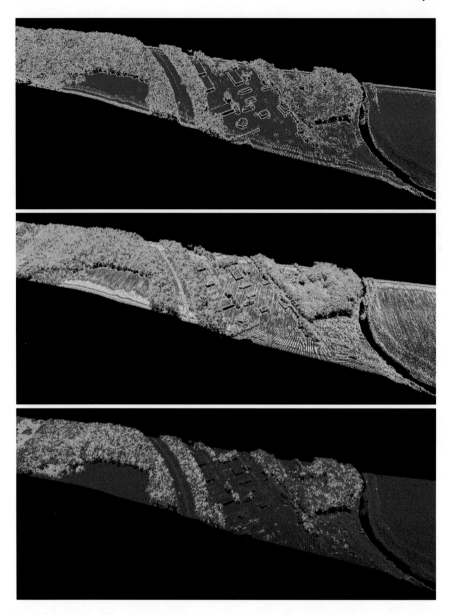

Fig. 6.3 Behavior of the dimensionality features of linearity L_λ (*top*), planarity P_λ (*center*), and scattering S_λ (*bottom*) for airborne laser scanning data when using neighborhoods formed by the $k = 25$ closest neighbors. The color encoding reaches from *violet* (representing a value of 0) via *blue*, *cyan*, *green*, and *yellow* to *red* (representing a value of 1)

Fig. 6.4 Behavior of the feature of planarity P_λ for terrestrial laser scanning data when using neighborhoods formed by the $k = 25$ closest neighbors. The color encoding reaches from *violet* (representing a value of 0) via *blue*, *cyan*, *green*, and *yellow* to *red* (representing a value of 1)

$$L_\lambda = \frac{e_1 - e_2}{e_1} \qquad (6.13)$$

$$P_\lambda = \frac{e_2 - e_3}{e_1} \qquad (6.14)$$

$$S_\lambda = \frac{e_3}{e_1} \qquad (6.15)$$

$$O_\lambda = \sqrt[3]{\prod_{i=1}^{3} e_i} \qquad (6.16)$$

$$A_\lambda = \frac{e_1 - e_3}{e_1} \qquad (6.17)$$

$$E_\lambda = -\sum_{i=1}^{3} e_i \ln(e_i) \qquad (6.18)$$

$$\Sigma_\lambda = \sum_{i=1}^{3} e_i \qquad (6.19)$$

$$C_\lambda = \frac{e_3}{\sum_{i=1}^{3} e_i} \qquad (6.20)$$

and they are nowadays commonly applied in LiDAR data processing.

6.3.2.2   2D Features

Besides a variety of 3D features, we may furthermore take into account that a 3D point cloud representing an observed scene does not provide a completely random point distribution since the 3D points represent the measured or derived counterpart of real object surfaces. Particularly in urban environments, we may face a variety of man-made objects which, in turn, are typically characterized by certain geometric constraints in terms of symmetry and orthogonality. In this regard, we take into account that such man-made objects often tend to provide (almost) perfectly vertical structures (e.g., building façades, walls, poles, traffic signs, or curbstone edges). For this reason, we may also involve geometric features resulting from a 2D projection of the 3D point cloud onto a horizontally oriented plane, i.e., the XY-plane according to our notation. Such *2D features* might reveal complementary information in comparison to the aforementioned 3D features and possibly also clear evidence about the presence of specific structures in the observed scene. Respective features may also be categorized into different groups.

Geometric 2D Properties Exploiting the 2D projection of a given 3D point $\mathbf{X}$ and its k closest neighbors, we may easily obtain basic geometric properties in analogy to the 3D case. Respective properties are, for instance, represented by the radius $r_{k\text{-NN,2D}}$ of the circular neighborhood defined by a 2D point and its k closest neighbors or the local point density D_{2D} [70]. In order to assess the geometric 2D properties corresponding to the locally optimized neighborhood, we again involve the closest neighbors based on 3D distances. Thus, the 2D projection of $\mathbf{X}$ and its k closest neighbors is used, and an additional nearest neighbor search in 2D space may be avoided.

Local 2D Shape Features Based on the introduced 2D projection, we may exploit the XY-coordinates of a 3D point $\mathbf{X}$ and its k closest neighbors in order to derive the *2D structure tensor* $\mathbf{S}_{2D}$ in analogy to the 3D structure tensor $\mathbf{S}_{3D}$. From its two eigenvalues $\lambda_{1,2D}$ and $\lambda_{2,2D}$ with $\lambda_{1,2D}, \lambda_{2,2D} \in \mathbb{R}$ and $\lambda_{1,2D} \geq \lambda_{2,2D} \geq 0$, the sum $\Sigma_{\lambda,2D}$ of eigenvalues and their ratio $R_{\lambda,2D}$ with

$$R_{\lambda,2D} = \frac{\lambda_{2,2D}}{\lambda_{1,2D}} \tag{6.21}$$

may be calculated and exploited as 2D features.

Features Based on a 2D Accumulation Map Interestingly, all the aforementioned geometric 2D features are based on the spherical neighborhood encapsulating a 3D point $\mathbf{X}$ and its k closest neighbors, where the neighborhood consists of $k = 10 \ldots 100$ neighbors (Sect. 6.3.1.3). Thus, the spherical neighborhood typically has a relatively small absolute size since we assume densely sampled 3D point cloud data. However, particularly man-made objects tend to provide a similar behavior across different height levels. Accordingly, we may also involve neighborhoods resulting from a *spatial binning*. In this regard, it has been proposed to introduce a second neighborhood definition by discretizing the 2D projection plane (i.e., the horizontally

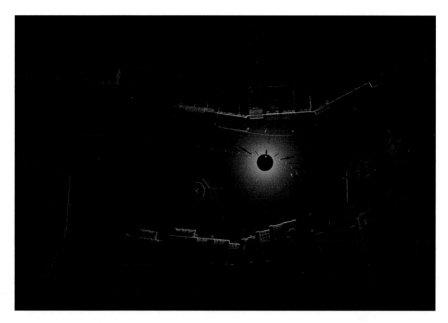

Fig. 6.5 Behavior of the number M of 3D points falling into the single bins of the 2D accumulation map: high values are indicated in *white*. Note that a consideration of this feature reveals building façades and, if the point density is high enough, we are also able to detect curbstones or poles in the scene. In this case, the feature M is even sufficient to recognize the scene depicted in Fig. 6.4

oriented XY-plane) and deriving a *2D accumulation map* with quadratic bins [83], e.g., with a side length of $0.20\ldots0.25$ m for MLS point cloud data. This allows us to also account for structures with larger extent in the vertical direction. Respective features assigned to the considered 3D point $\mathbf{X}$ arise from the number M of 3D points falling into the same bin as $\mathbf{X}$ (Fig. 6.5) as well as the maximum height difference ΔH and the standard deviation σ_H of height values of this bin.

6.3.3   Feature Selection

Particularly when dealing with many features or many training examples, using simpler and more efficient methods is advisable and, consequently, filter-based methods are often applied in order to select a subset of relevant features. Such filter-based methods evaluate simple and more intuitive relations between features and classes and possibly also among features (whereby the relations are based on well-known concepts of distance, information, dependency, or consistency), whereas those relations exploited by embedded methods are more sophisticated and hardly interpretable. When applying filter-based methods and thus representing relations in the form of score functions, one could argue that—in the sense of statistical learning or machine learning—embedded methods would be more appropriate, since the respective score function directly focuses on minimizing the classification error. However, involving an embedded method such as a Random Forest classifier [15] or an AdaBoost classifier [38] would directly introduce a dependency between selected features and the settings of a classifier, e.g., the number of involved weak learners, their type, and the (ideally high) number of considered choices per variable. In order to avoid an exhaustive classifier tuning and thus preserve applicability for nonexpert users, we focus on filter-based methods and we accept if these generally tend to provide a (slightly) weaker performance. Since we apply both univariate and multivariate filter-based methods, we briefly explain the basic ideas in the following subsections. For most of these techniques, however, it should be taken into account that balanced training data with an equal number of training examples per class is required in order to avoid a bias in feature selection.

6.3.3.1   Univariate Filter-Based Feature Selection

Generally, *univariate filter-based feature selection* methods rely on a score function which evaluates feature–class relations in order to discriminate between relevant and irrelevant features. More specifically, the involved score function evaluates the relation between the vector containing values of a single feature across all observations and the respective label vector. Thereby, the score function may address different intrinsic properties of the given training data such as distance, information, dependency, or consistency. Among a variety of score functions addressing a specific intrinsic property [30, 50, 153], the most popular ones are represented by simple metrics such as the Pearson correlation coefficient [100], the Gini index [43], the Fisher score [36], the information gain [105], or the symmetrical uncertainty [103]. Since some of these score functions are only defined for discrete-valued features, a respective discretization of continuous-valued features is typically introduced if required [33]. Thus, a variety of score functions allows to rank the extracted features according to their relevance.

6.3.3.2   Multivariate Filter-Based Feature Selection

Generally, *multivariate filter-based feature selection* methods rely on a score function which evaluates both feature–class and feature–feature relations in order to discriminate between relevant, irrelevant, and redundant features. A respective score function has been presented with ReliefF [63]. However, the involved score function may also be based on standard score functions applied for univariate filter-based feature selection. For instance, an approach taking the symmetrical uncertainty as correlation metric is represented by Correlation-based Feature Selection (CFS) [52]. Furthermore, the symmetrical uncertainty has also been used in the Fast Correlation-Based Filter (FCBF) [151]. However, there are also more complex approaches such as an approach relying on the minimal-redundancy-maximal-relevance (mRMR) criterion [101].

6.3.3.3 Involved Methods

In this subsection, we briefly explain two commonly applied approaches for univariate filter-based feature selection. Subsequently, we present a novel method relying on a general relevance metric composed of different (mostly univariate) filter-based feature selection methods. Furthermore, we consider three methods for multivariate filter-based feature selection.

Information Gain and Symmetrical Uncertainty In order to obtain an impression about the formulation of a score function, we briefly reflect two exemplary approaches. Defining random variables X for the feature values and C for the classes, we obtain the general definition of the Shannon entropy $E(X)$ indicating the distribution of feature values x_a as

$$E(X) = -\sum_a P(x_a) \ln P(x_a) \tag{6.22}$$

and the Shannon entropy $E(C)$ indicating the distribution of (semantic) classes c_b as

$$E(C) = -\sum_b P(c_b) \ln P(c_b) \tag{6.23}$$

respectively. From these, the joint Shannon entropy may be derived as

$$E(X, C) = -\sum_{a,b} P(x_a, c_b) \ln P(x_a, c_b) \tag{6.24}$$

which, in turn, is exploited in order to derive the *mutual information*

$$\begin{align}
\mathrm{MI}(X, C) &= E(X) + E(C) - E(X, C) \tag{6.25} \\
&= E(X) - E(X|C) \tag{6.26} \\
&= E(C) - E(C|X) \tag{6.27} \\
&= \mathrm{IG}(X|C) \tag{6.28} \\
&= \mathrm{IG}(C|X) \tag{6.29}
\end{align}$$

representing a symmetrical metric defined as *information gain* [105]. Accordingly, the amount of information gained about C after observing X is equal to the amount of information gained about X after observing C. A feature X is thus regarded as more correlated to the classes C than a feature Y if $\mathrm{IG}(C|X) > \mathrm{IG}(C|Y)$. For feature selection, the score function represented by information gain is evaluated independently for each feature, and those features with a high information gain are considered as relevant. Consequently, those features with the highest scores may be selected as relevant features. Note that information gain may also be derived via the conditional entropy, e.g., via $E(X|C)$ which quantifies the remaining uncertainty in X given that the value of the random variable C is known.

However, the score function represented by information gain is biased in favor of features with greater numbers of values since these appear to gain more information than others, even if they are not more informative [52]. This bias may be compensated by considering the metric

$$\text{SU}(X, C) = 2 \, \frac{\text{MI}(X, C)}{E(X) + E(C)} \tag{6.30}$$

defined as *symmetrical uncertainty* [103]. Note that the score functions represented by information gain and symmetrical uncertainty are only metrics for ranking features according to their relevance to the class and do not eliminate redundant features.

Our Approach Based on a General Relevance Metric With the intention to select a generally versatile set of relevant features which is not optimized with respect to a specific classifier, we focus on a classifier-independent solution which, in turn, results in both simplicity and efficiency. Consequently, we aim to involve a filter-based feature selection method. Such methods generally exploit a score function directly based on the training data, and the derived scores are used to rank the considered features according to their suitability. In this regard, the score function may address different intrinsic properties of the given training data such as distance, information, dependency, or consistency. Since the consideration of a single property may not be sufficient and thus different criteria of the data may be relevant [153], we define a general relevance metric which is based on several score functions.[1] More specifically, we involve $N_S = 7$ different score functions S_j which "evaluate" different intrinsic properties of the given training data:

1. The *Pearson correlation coefficient* $S_{\text{Pearson}} = S_1$ indicates to which degree a feature is correlated with the class labels [100]. A higher correlation results in higher values and thus indicates a higher relevance.
2. The *F-score* or *Fisher score* $S_{\text{Fisher}} = S_2$ represents the ratio between interclass and intraclass variance [36]. A larger value indicates that a feature is more likely to be discriminative.
3. The *information gain* $S_{\text{IG}} = S_3$ is a measure revealing the dependence between a feature and the class labels [105]. Higher values indicate more relevant features.
4. The *Gini index* $S_{\text{Gini}} = S_4$ provides a statistical measure of dispersion and thus an inequality measure which quantifies a feature's ability to distinguish between classes [43]. Smaller values of the Gini index indicate more relevant features.
5. The measure $S_{\chi^2} = S_5$ results from a χ^2-test which is used as a test of independence in order to assess whether the class labels are independent of a particular feature [103]. Following the provided implementation, a higher value indicates more relevance.
6. The measure $S_t = S_6$ results from applying a t-test on each feature and checking how effective it is for separating classes [103]. Following the provided implementation, a higher value indicates more relevance.

---

[1]Some of these score functions are also included in the *ASU Feature Selection Repository* and available at http://featureselection.asu.edu (last access: 30 May 2015).

7. The *ReliefF measure* $S_{\text{ReliefF}} = S_7$ indicates the contribution of a feature to the separation of samples from different classes [63]. If samples with the same label have close values and samples with different labels are well discriminated, this measure provides a higher value.

Note that these score functions address very different properties according to which feature subsets may be selected, and that the Gini index and the information gain are commonly used for node splitting in decision trees. In order to take into account several of these score functions, a combination of the respective ranking results is required. Consequently, for each score function S_j with $j = 1, \ldots, 7$, we derive a separate ranking of all features and denote the rank of a specific feature X_i given the score function S_j as $r(X_i|S_j)$. Thus, the rank $r(X_i|S_j)$ is an integer value in the interval $[1, N_F]$, where N_F denotes the number of involved features (i.e., $N_F = 21$ in our case). Smaller values for $r(X_i|S_j)$ reveal features with higher relevance when considering the respective score function S_j, whereas higher values for $r(X_i|S_j)$ reveal less suitable features. In order to obtain a general relevance metric $R(X_i)$ taking into account several score functions, we combine the separate ranking results across all score functions S_j by taking the mean rank $\bar{r}(X_i)$ of each feature X_i according to

$$\bar{r}(X_i) = \frac{1}{N_S} \sum_{j=1}^{N_S} r(X_i|S_j) \tag{6.31}$$

and we introduce a mapping to the interval $[0, 1]$ in order to interpret the result as relevance $R(X_i)$ of the feature X_i:

$$R(X_i) = 1 - \frac{\bar{r}(X_i) - 1}{N_F - 1} \tag{6.32}$$

In this regard, the *feature relevance* $R(X_i)$ represents a measure of feature importance and feature selection may be realized by only keeping a feature subset consisting of the few best-ranked features. In order to obtain the few best-ranked features, we may either start with a feature set consisting of only the best-ranked feature and successively add the next best-ranked feature to the feature set (and thus conduct Sequential Forward Selection), or start with the full feature set and successively remove the worst-ranked feature from the feature set (and thus conduct Sequential Backward Elimination).

Correlation-based Feature Selection (CFS) An approach for selecting relevant features and thereby reducing redundancy has been presented with *Correlation-based Feature Selection* (CFS) [52] which relies on using the symmetrical uncertainty (and thus a standard score function applied for univariate filter-based feature selection) as correlation metric. Considering random variables X_i for the features and C for the class labels and furthermore defining $\bar{\rho}_{XC}$ as the average correlation between features and classes as well as $\bar{\rho}_{XX}$ as the average correlation between different features, the relevance R of a feature subset consisting of $n \leq N_F$ features results in

$$R(X_{1...n}, C) = \frac{n\bar{\rho}_{XC}}{\sqrt{n + n(n-1)\bar{\rho}_{XX}}} \tag{6.33}$$

which is to be maximized by searching the feature subset space [52], i.e., by iteratively adding a feature to the feature subset (forward selection) or removing a feature from the feature subset (backward elimination) until R converges to a stable value.

Fast Correlation-Based Filter (FCBF) For comparison only, we also consider feature selection based on a *Fast Correlation-Based Filter (FCBF)* [151] which involves heuristics and thus does not meet our intention of a fully generic methodology. In order to decide whether features are relevant to the class or not, a typical feature ranking based on the symmetrical uncertainty is conducted which yields the feature-class correlation. If the symmetrical uncertainty is above a certain threshold, the respective feature is considered to be relevant. In order to decide whether a relevant feature is redundant or not, the symmetrical uncertainty among features is compared to the symmetrical uncertainty between features and classes which, in turn, allows to remove redundant features and only keep predominant ones.

Minimal-Redundancy-Maximal-Relevance (mRMR) criterion Furthermore, we consider an approach addressing the aims of both univariate and multivariate filter-based feature selection methods. Note that, in this regard, univariate filter-based methods focus on selecting the best-ranked features with the highest relevance (i.e., maximal relevance selection), whereas multivariate filter-based methods focus on selecting features with the minimal redundancy (i.e., minimal redundancy selection). In order to account for both aspects, an approach combining two constraints for (i) minimal redundancy selection and (ii) maximal relevance selection has been presented with the *minimal-redundancy-maximal-relevance (mRMR) criterion* [101]. However, one limitation of this approach consists of the fact that the optimal number of features in the final feature set has to be determined which may be done either heuristically or by involving a classifier (and thus obtaining a wrapper-based method). Since we focus on efficiency and therefore do not want to apply a classifier-dependent feature selection, the method of choice involves heuristics and we select a feature set consisting of 10 features as proposed in [101].

6.3.4 Classification

Finally, the derived feature vectors consisting of either all or only relevant features are provided to a classifier which returns a respective assignment to one of the specified (semantic) classes. In this regard, a *supervised classification* scheme is typically involved which is based on the idea of exploiting given training data in order to train a classifier which, after the training phase, is able to generalize to new unseen data. Introducing a formal description, the training data may be represented by a training set $\mathcal{X} = \{(\mathbf{f}_i, c_i)\}$ with $i = 1, \ldots, N_{\mathcal{X}}$ consisting of $N_{\mathcal{X}}$ training examples $(\mathbf{f}_i, c_i)$ which, in turn, consist of an assignment between a feature vector $\mathbf{f}_i$ in a

D-dimensional feature space and a respective class label $c_i \in \{1, \ldots, N_C\}$, where N_C represents the number of classes. In contrast, the test data may be represented as test set $\mathscr{Y} = \{\mathbf{f}_j\}$ with $j = 1, \ldots, N_{\mathscr{Y}}$ which only consists of $N_{\mathscr{Y}}$ feature vectors $\mathbf{f}_j \in \mathbb{R}^D$. If available, the respective class labels $c_j \in \{1, \ldots, N_C\}$ may be used for evaluation.

For multiclass classification, we involve different classifiers and thereby focus on applicability in terms of simplicity, efficiency, and reproducibility. For this reason, we follow the strategy of individual classification, where each 3D point is classified by only exploiting the respective feature vector and many respective approaches are available in a variety of software tools. Since quite different learning principles may be involved in order to infer a function between feature vectors and class labels in the training phase and since we intend to obtain general conclusions about our scientific contribution consisting of the use of features with increased distinctiveness due to the consideration of individually optimized neighborhoods, we involve a variety of classifiers belonging to different categories and briefly present their main ideas in the following subsections.

For all learning principles, we take into account that an unbalanced distribution of training examples per class in the training set may have a detrimental effect on the training process [23, 27]. In order to avoid such an effect, we introduce a *class re-balancing* by randomly sampling the same number of training examples per class which yields a reduced training set.[2] Thus, end-users will not only get an impression about the performance of single approaches, but also a comprehensive comparison.

6.3.4.1 Instance-Based Learning

Instance-based learning relies on directly comparing unseen feature vectors to those feature vectors in the training set and does therefore neither require parameter estimation nor the assumption of a certain model. The only requirement is the definition of a similarity metric upon which the comparison may be carried out. This similarity metric may for instance be based on the Euclidean distance, a general Minkowski metric or other distance metrics. Thus, a simple and straightforward approach for instance-based learning is represented by a *Nearest Neighbor (NN) classifier* which assigns each feature vector the class label of the most similar training example. A generalization of the NN classifier is represented by a *k Nearest Neighbor (k-NN) classifier* [26] which, for each feature vector, selects the k closest training examples and classifies according to the majority vote of their class labels. Since the induction (i.e., the generalization) is delayed to the classification of unseen feature vectors, instance-based learning techniques are also referred to as *lazy-learning techniques* [77].

---

[2]Note that this might result in a duplication of training examples if the selected number is larger than the number of training examples belonging to the smallest class.

6.3.4.2   Rule-Based Learning

Rule-based learning relies on a knowledge representation in terms of (mostly binary) decisions. As most prominent example, *Decision Tree (DT) classifiers* conduct a series of simple tests which, in turn, are organized hierarchically in a tree structure [105]. The construction of a decision tree typically relies on a top-down strategy, where at each step that variable is chosen which best splits the given training data. Thereby, the recursive partitioning strongly depends on the definition of the split function and the respective stopping criterion. For both criteria, we use standard settings.

6.3.4.3   Probabilistic Learning

Probabilistic learning relies on the idea of deriving an explicit underlying probabilistic model and inferring the most probable class label for each observed feature vector. A respective approach has been presented with the *Naïve Bayesian (NB) classifier* [56] which represents a probabilistic classifier based on Bayes' theorem and the naïve assumption of all features being conditionally independent. In the training phase, a set of class probabilities and conditional probabilities for the occurrence of a class given a specific class label have to be determined based on the training set $\mathscr{X}$. Thus, in the classification process, a new feature vector of a test set $\mathscr{Y}$ may be assigned the most likely class label by simply applying Bayes' theorem. However, since conditional independence is assumed, correlated features cannot be modeled appropriately. Hence, an alternative is represented by a classical Maximum Likelihood (ML) classifier which may be derived by considering the distribution-based *Bayesian Discriminant Analysis*. Thereby, in the training phase, a multivariate Gaussian distribution is fitted to the given training data, i.e., the parameters of a Gaussian distribution are estimated for each class by parameter fitting. For a *Linear Discriminant Analysis (LDA) classifier*, the same covariance matrix is assumed for each class and therefore only the means may vary whereas, for a *Quadratic Discriminant Analysis (QDA) classifier*, the covariance matrix of each class may also vary. Thus, in order to classify a new feature vector, the respective probabilities that the considered feature vector belongs to the different classes are evaluated, and the label of the class with the maximum probability is assigned.

6.3.4.4   Max-Margin Learning

Max-margin learning relies on maximizing the distance between samples of different classes in the feature space. The most prominent approach in this regard has been presented with a *Support Vector Machine (SVM) classifier* [25]. Originally, an SVM represents a binary classifier which is trained to linearly separate two classes by constructing a hyperplane or a set of hyperplanes in a high-dimensional feature space. Such a linear separation is however not possible for a wide range of classification problems and, for this reason, a kernel function is typically introduced which implicitly maps the training data into a new feature space of higher dimensionality where the data is linearly separable. An SVM classifier addressing multiclass classification is derived by combining several binary SVMs, and we use a respective implementation provided in the LIBSVM package [21], where the SVM classifier is based on a one-against-one approach and a (Gaussian) radial basis function (RBF) as kernel. Accordingly, for each pair of classes, an SVM is trained to distinguish samples of one class from samples of the other class. Such a principle may allow a better training and subsequent discrimination of classes closely located in the feature space when considering general characteristics of TLS and MLS data acquired in urban environments which, in turn, typically contain many objects with similar shapes or at least similar geometrical behavior (e.g., poles, wires, trunks, traffic signs or traffic lights). While a *hard-margin* approach aims at maximizing the margin and thereby correctly classifying all training data (which typically results in very complex decision boundaries), it does not take into account that both of these objectives may possibly be conflicting and a loss with respect to the generalization capability might thus occur. In contrast, a *soft-margin* approach also allows for a certain amount of misclassifications on the training data (which typically results in less complex decision boundaries) and thus often provides the desired capability of generalization. In order to define a trade-off between the complexity of decision boundaries and classification errors on the training data, a respective term is added to the SVM formulation and weighted with a parameter C penalizing classification errors on the training data. As a consequence, the classification results of an SVM classifier strongly depend on two parameters represented by (i) the parameter γ representing the width of the RBF kernel and (ii) the parameter C penalizing classification errors. In order to select suitable values, we follow the standard procedure and therefore conduct a grid search in an appropriate subspace (γ, C). For more details, we refer to [8, 12, 18, 21, 25, 32].

6.3.4.5 Ensemble Learning

Ensemble learning relies on the idea of strategically generating a set of weak learners and combining them in order to create a single strong learner. For this purpose, a rather intuitive strategy is represented by *bagging* [14], where a predefined number of weak learners of the same type is trained independently from each other on bootstrapped replica of the training data (i.e., subsets of the training data which are randomly drawn with replacement [31]). Thus, the weak learners are all randomly different from one another and a decorrelation between individual hypotheses may be expected which, in turn, results in an improved generalization and robustness when taking the respective majority vote over all hypotheses [27]. A very popular approach in this regard is represented by a *Random Forest (RF) classifier* [15] which relies on decision trees as weak learners. By considering simplified decision trees in the form of *ferns*, a modification in terms of a nonhierarchical structure consisting of a set of ferns as weak learners whose hypotheses are combined in a Naïve Bayesian way has been presented with a *Random Fern (RFe) classifier* [96].

For the respective Random Forest classifier involved in our experiments, the settings (i.e., the number N_T of decision trees, the maximum tree depth d_{max} of each decision tree, the minimum allowable number n_{min} of training examples for a tree node to be split, the number n_a of active variables to be used for the test in each tree node, etc.) have been determined via various tests based on different choices. Accordingly, we use 100 decision trees for the Random Forest classifier. In a similar way, we have conducted tests in order to determine the settings for the Random Fern classifier (i.e., the number N_T of ferns, the fern depth d_T, etc.). Based on these tests, we select 100 ferns for the Random Fern classifier.

A different strategy for strategically generating weak learners and combining these to a strong learner has been presented with *boosting* [113] which relies on incrementally generating a set of weak learners over consecutive iterations and different distributions of the training data. In each iteration, a subset of the training data is selected in order to train a weak learner and thus get a weak hypothesis with low error with respect to the true labels. After a certain number of iterations, all hypotheses are combined by taking the respective majority vote over all hypotheses. Since boosting has originally been proposed for binary classification problems, an extension to multiclass classification has been presented with Adaptive Boosting [38] which is commonly referred to as *AdaBoost*. As a result of different tests for selecting suitable settings, we use an *AdaBoost (AB) classifier* based on 100 decision trees as weak learners.

6.3.4.6   Deep Learning

Deep learning relies on biological observations and has been inspired by biological neural networks such as the human brain which consists of a huge number of connected neurons and provides the capability to model high-level abstractions in given data. The transfer from biological neural networks to artificial neural networks requires the definition of a model for the signal processing in an artificial neuron and the definition of a topology specifying the arrangement of neurons in the network. While a variety of topologies have been investigated, the most prominent topology is still represented by the *Multi-Layer Perceptron* (*MLP*) which consists of multiple layers of neurons: an input layer, one or two hidden layers and an output layer. Each layer is fully connected to the next one, whereby each connection is characterized by a weight factor. Thus, a number of weighted inputs is provided to each neuron which, in turn, maps these inputs to its output via an activation function.

Generally, the number of neurons for both input and output layer is derived via the respective training examples, where we may for instance exploit one neuron per feature (i.e., one neuron per entry in the feature vector) in the input layer and one neuron for each possible outcome (i.e., one neuron per defined class) in the output layer. In contrast, a suitable number of neurons in the hidden layer has to be determined heuristically and it has to be taken into account that relatively high numbers tend to result in an overfitting, whereas relatively low numbers tend to result in an insufficient predictive accuracy. Once the number of neurons in the hidden layer has been specified, the training phase typically focuses on estimating all weight factors via *backpropagation* [107] which represents a gradient descent technique for minimizing an error function in a high-dimensional space based on the given training data. Subsequently, unseen data may be provided as input for the network which propagates this data from layer to layer and, finally, to a respective output (label).

In order to determine useful settings, we have conducted various tests and have thereby considered different choices for (i) the number of neurons in the hidden layer, (ii) the definition of the activation functions and (iii) the training procedure itself. Based on these tests, we select a Multi-Layer Perceptron with 11 neurons in the hidden layer, linear activation functions for neurons in the input or output layer, activation functions in the form of logistic sigmoid functions for neurons in the hidden layer and a training based on the Resilient Backpropagation algorithm [106] for our experiments.

6.4 Extension Toward Large-Scale 3D Scene Analysis

While we expect that using individual neighborhoods of optimal size has a significantly beneficial impact on the classification results, we should not forget that such an optimal neighborhood size selection causes a drastic increase in computational effort, since each 3D point is treated individually. As a consequence, the described framework will probably be capable to process 3D point clouds with only a few millions of 3D points without further adaptations. When considering significantly larger 3D point clouds with possibly billions of 3D points—which is the aim of recent effort in order to obtain an adequate 3D model of a whole city like Paris [98, 121]—an extension of the presented framework toward data-intensive processing has to be introduced.

In the scope of our work, an appropriate extension has to satisfy specific constraints. The first constraint is based on the fact that the extension of the framework solely addresses the scalability of the involved methods in order to process larger datasets without affecting the quality of the results derived for 3D scene analysis. The second constraint is represented by the fact that the extension of framework only considers geometric aspects based on the given TLS/MLS point cloud data, since we only exploit geometric features extracted from densely sampled 3D point clouds for 3D scene analysis. Satisfying these two constraints, we focus on a successive processing of huge 3D point cloud data by shifting a 2D sliding window function within a horizontally oriented plane (i.e., the XY-plane according to our notation) in discrete steps and thereby involving a small padding region in order to avoid discontinuities at its borders. Subsequently, the derived partial results are merged together in order to obtain the respective results for the whole scene. Thus, the extension of our framework is represented by an initial partitioning of the scene into subparts which, in turn, are extended by small padding regions at the borders and may be processed in parallel.

It becomes obvious that different settings have to be selected for such a partitioning of the scene into subparts. On the one hand, the size of the sliding window should be chosen in a way that, for each discrete step, a still reasonable number of 3D points is in consideration, since a small size of the sliding window results in lots of padding regions and a lot of partial results which have to be merged, whereas the maximum size of the sliding window is practically limited by the memory available for data processing. On the other hand, the size of the padding region should be chosen in a way that discontinuities at the borders of the sliding window may be suppressed. Furthermore, the shape of the sliding window itself has to be selected in a suitable way, where we focus on a rectangular window aligned with respect to the horizontal 2D coordinates of the considered 3D points in order to efficiently partition a scene. For the sake of simplicity, we focus on two specific scenarios and may select the one which is suited best for the given data:

- For general scenes, we propose the use of a *tiling approach*, where the scene is divided into single tiles representing data within a well-defined, rectangular area in the horizontally oriented plane. Thus, a successive processing of single tiles is possible and therefore also the analysis of huge 3D point clouds at city scale, where directly applying our main framework is intractable due to the computational burden with respect to processing time and memory consumption.
- Without loss of generality, we may also take into account that the recently published benchmark datasets represent straight street sections with a length of about $160 \ldots 200$ m [98, 121]. Consequently, the tiling approach may slightly be modified to a *slicing approach*, where the single slices are oriented orthogonally with respect to the street direction and have a specified width along the street direction while exhibiting an infinite extent along the two perpendicular directions.

Note that, for both *scene partitioning* schemes, those 3D points within the small padding around the considered tile or slice are only used if they are within the local neighborhood of any 3D point located in the considered part of the scene in order to suppress artifacts at boundaries between tiles or slices. Due to the high point density of recent LiDAR point cloud datasets and some empirical tests, we consider a tile in the form of a small quadratic area of 10 m × 10 m, a slice with a width of 10 m and a padding with a width of 0.50 m as sufficient. As a consequence, we may apply our main framework for each tile or for each slice. Thus, large-scale 3D scene analysis is composed of successive steps which exploit the same methodology and could therefore be parallelized if respective hardware is available.

For the sake of applicability, we should also keep in mind that, generally, an alternative to scene partitioning is represented by *streaming methods* which may handle extremely large 3D point cloud datasets without requiring a special treatment for boundaries of tiles [155]. More specifically, such a streaming method takes data as a stream (e.g., in the form of disk files) and relies on the assumption that most computations only require data access within a small local area and are therefore insensitive to global variables. An appropriate splitting into separate disk files is however not available for the currently available benchmark datasets which we intend to involve in order to evaluate the performance of our framework.

6.5 Extension by Involving Contextual Information

Finally, it should be taken into account that approaches for 3D point cloud classification typically consider the different components of a typical processing workflow (i.e., neighborhood selection, feature extraction, and classification) independently from each other. However, it would seem desirable to connect these components by sharing the results of crucial tasks across all of them. Such a connection would not only be relevant for the interrelated issues of neighborhood selection and feature extraction, but also for the question of how to involve spatial context in the classification step. Accordingly, we also consider an extension of our main framework by involving contextual information in order to verify our scientific contribution consisting of the use of features with increased distinctiveness due to the consideration of individually optimized 3D neighborhoods. This extension relies on involving individually optimized 3D neighborhoods for both the extraction of distinctive geometric features and the contextual classification of 3D point cloud data.

Such an extension may principally be motivated by the fact that a respective combination provides further important insights into the interrelated issues of neighborhood selection, feature extraction and contextual classification. On the one hand, we claim that using features extracted from individual neighborhoods has a significantly beneficial impact on the individual classification of 3D points. On the other hand, using contextual information might even have more influence on the classification accuracy, since it takes into account that class labels of neighboring 3D points tend to be correlated. This raises the question whether the use of features extracted from neighborhoods of individual size still improves the classification accuracy when contextual classification is applied, and whether it is beneficial to use the same neighborhood definition for contextual classification. More specifically, contextual learning not only relies on the respective feature vectors but also on relationships among 3D points in a local neighborhood, whereby the relationships have to be inferred from the training data (see Sect. 6.2.4). For this purpose, statistical models of context are typically involved which are based on interactions between neighboring point pairs and, consequently, the considerations made about the size of a local neighborhood (see Sects. 6.2.1 and 6.3.1) may also apply for the selection of the set of 3D points interacting with a given 3D point. In this regard, we explicitly want to point out that existing investigations are usually based on using a fixed radius or on considering a fixed number k of nearest neighbors either in 2D or in 3D space in order to specify the respective neighborhoods. However, recent investigations on the impact of varying the radius of a cylindrical neighborhood for defining the set of neighbors [90] clearly indicate a saturation effect when increasing the radius so that, in the respective experiments, the number of involved neighbors is set to $k = 7$ for all 3D points of the considered 3D point cloud.

Since *Conditional Random Fields* (*CRFs*) [64, 67] represent one of the most popular approach for contextual learning, we involve a CRF in our experiments and intend to investigate the effect of using individual 3D neighborhoods of optimal size for defining the edges of the considered CRF. Accordingly, we consider different neighborhood definitions as the basis for feature extraction, use a CRF for contextual classification, and compare the respective classification results with those obtained when using a Random Forest classifier [15]. Since we may also define the unary terms (i.e., the association potentials) of a CRF based on a Random Forest classifier, we may thus quantify the influence of the context model on the classification results. In order to allow a better understanding of how our definition of individual 3D neighborhoods affects a CRF, we explain the most important ideas in this regard in the following.

Generally, a CRF is an undirected graphical model that allows to model interactions between neighboring objects to be classified and thus to model local context. The underlying graph $G(n, e)$ consists of a set of nodes n and a set of edges e, where the latter are responsible for the context model. In our case, similarly to [92], the nodes $n_i \in n$ correspond to the 3D points $\mathbf{X}_i$ of the considered 3D point cloud, whereas the edges $e_{ij} \in e$ connect neighboring pairs of nodes (n_i, n_j). Consequently, the number of nodes in the graph coincides with the number N_P of 3D points $\mathbf{X}_i$ to be classified and the classification, in turn, aims at assigning a class label $c_i \in \{c^1, \ldots, c^{N_C}\}$ to each 3D point $\mathbf{X}_i$ (and thus to each node n_i of the graph), where N_C is the number of classes, superscripts indicate specific class labels corresponding to an object type, and subscripts indicate the class label of a given 3D point. Due to the mutual dependencies between the class labels at neighboring 3D points induced by the edges of the graph, the class labels of all 3D points have to be determined simultaneously. Representing the class labels of all 3D points in a vector $\mathbf{C} = \left[c_1, \ldots, c_i, \ldots, c_{N_P}\right]^T$ and furthermore denoting the combination of all input data by $\boldsymbol{\xi}$, we intend to determine the configuration $\mathbf{C}$ of class labels which maximizes the posterior probability $p(\mathbf{C}|\boldsymbol{\xi})$ given by

$$p(\mathbf{C}|\boldsymbol{\xi}) = \frac{1}{Z(\boldsymbol{\xi})} \left(\prod_{i \in n} \phi(\boldsymbol{\xi}, c_i) \prod_{i \in n} \prod_{j \in \mathcal{N}_i} \psi(\boldsymbol{\xi}, c_i, c_j) \right) \qquad (6.34)$$

where $Z(\boldsymbol{\xi})$ represents a normalization constant called the *partition function* [64]. Since this partition function does not depend on the class labels, it may be neglected in the classification step. Among the further terms in the provided equation, the functions $\phi(\boldsymbol{\xi}, c_i)$ are referred to as *association potentials* which provide local links between the data $\boldsymbol{\xi}$ and the local class labels c_i, whereas the functions $\psi(\boldsymbol{\xi}, c_i, c_j)$ are referred to as *interaction potentials* which are responsible for the local context model and provide the links between the class labels (c_i, c_j) of the pair of nodes connected by the edge e_{ij} and the data $\boldsymbol{\xi}$. Thus, the association potential representing a unary potential indicates how a specific 3D point $\mathbf{X}_i$ would be labeled if we only consider the respective feature vector, whereas the interaction potential representing a pair-wise potential indicates how a specific 3D point $\mathbf{X}_i$ would be labeled if we consider other feature vectors and labels as well. Furthermore, $\mathcal{N}_i$ denotes the set of neighbors of node n_i that are linked to n_i by an edge. Details about our definitions of the individual terms and the local neighborhood are given in the following subsections.

6.5.1 Association Potentials

Generally, any local discriminative classifier whose output can be interpreted in a probabilistic way may be used to define the *association potentials* $\phi(\xi, c_i)$ in Eq. (6.34). Note that, since the data ξ appears without an index, the association potential for node n_i may principally depend on all the data [64]. This is usually considered by defining site-wise feature vectors $\mathbf{f}_i(\xi)$, i.e., in our case one such vector per 3D point $\mathbf{X}_i$ to be classified. Consequently, we use the feature vectors $\mathbf{f}_i$ defined as site-wise vectors $\mathbf{f}_i(\xi)$ whose components are functions of the data within a neighborhood of the 3D point $\mathbf{X}_i$. In our experiments, we will compare different variants of these feature vectors based on different definitions of the local neighborhood used for computing the features as defined in Sect. 6.3.1. The association potential may be defined as the posterior probability of a local discriminative classifier based on $\mathbf{f}_i(\xi)$ [64]:

$$\phi(\xi, c_i) = p\left(c_i | \mathbf{f}_i(\xi)\right) \qquad (6.35)$$

For individual classification, a good trade-off between classification accuracy and computational effort may be achieved by using a Random Forest classifier [15], where each involved decision tree casts a vote for one of the class labels c^l. In order to use the output of a Random Forest classifier for the association potential, we define the posterior probability of each class label c^l to be the ratio of the number N_l of votes cast for that class and the number N_T of involved decision trees:

$$p\left(c_i = c^l | \mathbf{f}_i(\xi)\right) = \frac{N_l}{N_T} \qquad (6.36)$$

6.5.2 Interaction Potentials

In a similar way as the association potentials, the *interaction potentials* may also be based on the output of a discriminative classifier [64]. In recent investigations [92], for instance, a Random Forest classifier [15] is used as discriminative classifier delivering a posterior probability $p\left(c_i, c_j | \mu_{ij}(\xi)\right)$ for the occurrence of the class labels (c_i, c_j) at two neighboring 3D points $\mathbf{X}_i$ and $\mathbf{X}_j$ given an observed interaction feature vector μ_{ij} (the respectively concatenated node feature vectors). Thus, the interaction potentials are defined as $\psi(\xi, c_i, c_j) = p\left(c_i, c_j | \mu_{ij}(\xi)\right)$, and the respectively derived results reveal that such a model delivers a better classification performance for classes having a relatively small number of instances in a 3D point cloud. In order to apply such an approach, however, a sufficient number of training examples is required for each type of class transition. For instance, if the number of classes is N_C, enough training examples for $N_C \times N_C$ such transitions would be required, which may be prohibitive. As a consequence, we focus on the use of a simpler model representing a variant of the *contrast-sensitive Potts model* [11] in order to define the interaction potentials:

$$
\log\left(\psi\left(\boldsymbol{\xi}, c_i, c_j\right)\right) = \delta_{c_i c_j} \cdot w_1 \cdot \frac{N_a}{N_{k_i}} \cdot \left(w_2 + (1 - w_2) \cdot e^{-\frac{d_{ij}^2(\boldsymbol{\xi})}{2 \cdot \sigma^2}}\right) \tag{6.37}
$$

In this equation, $d_{ij}^2\left(\boldsymbol{\xi}\right) = \left\|\mathbf{f}_i\left(\boldsymbol{\xi}\right) - \mathbf{f}_j\left(\boldsymbol{\xi}\right)\right\|^2$ represents the square of the Euclidean distance between the node feature vectors $\mathbf{f}_i\left(\boldsymbol{\xi}\right)$ and $\mathbf{f}_j\left(\boldsymbol{\xi}\right)$ of the two nodes connected by the edge e_{ij}. Furthermore, $\delta_{c_i c_j}$ represents the Kronecker delta returning 1 if the class labels c_i and c_j are identical and 0 otherwise. The parameter σ denotes the average square distance between the feature vectors at neighboring training points, N_a represents the average number of edges connected to a node in the CRF and N_{k_i} is the number of neighbors of node n_i. The remaining variables w_1 and w_2 represent weight parameters, whereby the weight parameter w_1 influences the impact of the interaction potential on the classification results. Note that the normalization of the interaction potential by the ratio N_a/N_{k_i} is required for the interaction potentials to have an equal total impact on the classification of all nodes [137]. The model presented in Eq. (6.37) will result in a data-dependent smoothing of the classification results, and the second weight parameter $w_2 \in [0, 1]$ describes the degree to which smoothing will depend on the data.

6.5.3   *Definition of the Neighborhood*

Considering the provided definitions, it becomes apparent that an important question in the application of a CRF consists of an appropriate definition of the neighborhood $\mathcal{N}_i$ for each node n_i corresponding to a 3D point $\mathbf{X}_i$. While, for images, we may for instance use the four direct neighbors of a pixel on the discrete image raster [64], such a simple definition is impossible when considering 3D point cloud data. For the latter, the definition of the local neighborhood $\mathcal{N}_i$ is typically based on a spherical neighborhood definition with fixed radius, a cylindrical neighborhood definition with fixed radius or a neighborhood formed by a fixed number k of closest neighbors (see Sect. 6.2.1). Thereby, the choice of a suitable neighborhood definition typically depends on the given 3D point cloud data. For 3D point clouds acquired via airborne laser scanning, it has been shown that a cylindrical neighborhood is to be preferred as, for instance in an urban area, building façades will only be represented by a relatively small number of 3D points, while the height differences between neighboring points (in 2D) carry a lot of information [92]. In the scope of our work, we focus on data acquired with either terrestrial or mobile laser scanners and, consequently, we may expect a relatively high number of 3D points on building façades so that applying a cylindrical neighborhood does not make much sense. Hence, we consider the k

closest neighbors of each 3D point (based on 3D distances) in order to define the edges of the graph. As already explained in Sects. 6.2.1 and 6.3.1, the selection of an identical scale parameter k across all 3D points of a considered point cloud may not be appropriate. For this reason, we define the neighborhood size as explained in Sect. 6.3.1 in order to obtain *spatially varying definitions* of the local neighborhood. In this regard, we take into account that, due to performance reasons, we have to apply stricter limits to the size of the local neighborhood than for the size of the local neighborhood used to extract the features. Consequently, if the neighborhood size determined according to one of the methods defined in Sects. 6.2.1 and 6.3.1 is larger than a threshold $k_{max,CRF}$, we will select a maximum neighborhood size of $k_{max,CRF}$ closest neighbors. In our experiments, we will involve several definitions of the neighborhood size and some of these definitions also rely on the use of a fixed scale parameter k. For those variants with a variable scale parameter k, the average number N_a of neighbors in Eq. (6.37) will only be based on the actual number of neighbors per node (i.e., after enforcing the threshold $k_{max,CRF}$).

6.5.4 Training and Inference

In analogy to other supervised classification schemes, we need training data, i.e., a set of 3D points with known class labels, in order to determine the parameters of a CRF. Based on the training data, the parameters of the two types of potentials represented by the association potentials and the interaction potentials are trained independently from each other. For the association potentials, this involves the training of a discriminative classifier which is represented by a Random Forest classifier [15] in our case, and we randomly select an identical number of training examples per class which, in turn, is required as otherwise a class with many examples might lead to a bias toward that class in training [23, 27]. For the interaction potentials, the parameter σ is determined as the average square distance between neighboring points in the training data based the same local neighborhood that is used for the definition of the graph in classification. Furthermore, the weight parameters w_1 and w_2 which could generally be selected based on a technique such as cross validation [126] are set to values which have been derived from empirical tests.

For *inference*, i.e., for the determination of the label configuration $\mathbf{C}$ maximizing the posterior probability $p(\mathbf{C}|\xi)$ in Eq. (6.34) once the parameters of the potentials are known, we have to take into account that, for a large graph with cycles, a strict evaluation of $p(\mathbf{C}|\xi)$ and thus exact inference may be computationally intractable due to mutual dependencies between the random variables [92]. Hence, approximate methods have to be applied and we therefore use Loopy Belief Propagation [40], a standard optimization technique for graphs with cycles.

6.6   Experimental Results

In order to test the performance of our framework, we involve publicly available benchmark datasets (Sect. 6.6.1) and conduct experiments focusing on different aspects (Sect. 6.6.2). The achieved results are presented in the form of an extensive evaluation (Sect. 6.6.3), where we provide the derived results for optimal neighborhood size selection in comparison to standard neighborhood definitions while focusing on a comparison of single approaches for classification and feature selection.

6.6.1   Datasets

While focusing on (large-scale) 3D scene analysis, we explicitly intend to facilitate an objective comparison between our results and those results of other recent and future methodologies. Hence, in order to evaluate the performance of our framework, we involve three publicly available benchmark datasets consisting of 3D point cloud data and respective point-wise annotations. All these datasets represent densely sampled urban environments acquired with a mobile laser scanning (MLS) system. More details are provided in the following subsections.

6.6.1.1   Oakland 3D Point Cloud Dataset

One of the most widely used MLS datasets has been released with the *Oakland 3D Point Cloud Dataset*[3] [85, 86] which represents an urban environment and has been acquired in the vicinity of the CMU campus in Oakland, USA. For data acquisition, a mobile platform was used which has been equipped with side looking SICK LMS laser scanners used in push-broom mode. In order to allow a fair comparison of approaches, a separation of the dataset into a training set $\mathscr{X}_0$, a validation set $\mathscr{V}$, and a test set $\mathscr{Y}$ is provided (see Table 6.3), where each 3D point is assigned one of the five semantic labels *wire, pole/trunk, façade, ground*, and *vegetation* as visualized in Fig. 6.6.

Since an unbalanced distribution of training examples across all classes (as given in the training set $\mathscr{X}_0$) may have a detrimental effect on feature selection methods as well as the training process [27], we conduct a class re-balancing by randomly selecting the same number of 1,000 training examples per class in order to obtain a balanced training set $\mathscr{X}$, while the original test set $\mathscr{Y}$ is used for evaluation.

---

[3]The Oakland 3D Point Cloud Dataset with five semantic classes is publicly available at http://www.cs.cmu.edu/~vmr/datasets/oakland_3d/cvpr09/doc/ (last access: 30 May 2015).

Table 6.3 Number of samples per class for the Oakland 3D Point Cloud Dataset: note that the training set $\mathscr{X}_0$ and the test set $\mathscr{Y}$ are already given, while the training set $\mathscr{X}$ has been derived via class re-balancing

Class	Training set $\mathscr{X}_0$	Training set $\mathscr{X}$	Test set $\mathscr{Y}$
Wire	2,571	1,000	3,794
Pole/trunk	1,086	1,000	7,933
Façade	4,713	1,000	111,112
Ground	14,121	1,000	934,146
Vegetation	14,441	1,000	267,325
Σ_{set}	36,932	5,000	1,324,310

Fig. 6.6 3D point cloud of the Oakland 3D Point Cloud Dataset where the color encoding addresses the five considered semantic classes (*wire: blue; pole/trunk: red; façade: gray; ground: brown; vegetation: green*)

6.6.1.2   Paris-Rue-Madame Database

The *Paris-rue-Madame database*[4] [121] represents a 3D point cloud dataset which has been acquired in February 2013 in the city of Paris, France. For data acquisition, the mobile laser scanning system L3D2 [45] was used which captures information about the local 3D geometry with a Velodyne HDL32. The acquired dataset consists of 20M 3D points corresponding to a digitized street section with a length of about 160 m, and a respective annotation has been conducted in a manually assisted way in order to provide a ground truth with respect to point-wise labels (26 different classes) and segmented objects (642 objects in total). This annotation relies on an

---

[4] *Paris-rue-Madame database: MINES ParisTech 3D mobile laser scanner dataset from Madame street in Paris.* © 2014 MINES ParisTech. MINES ParisTech created this special set of 3D MLS data for the purpose of detection-segmentation-classification research activities, but does not endorse the way they are used in this project or the conclusions put forward. The database is publicly available at http://cmm.ensmp.fr/~serna/rueMadameDataset.html (last access: 30 May 2015).

initial segmentation based on elevation images [120] which is followed by a manual refinement [121].

For our experiments, we discard 3D points belonging to classes which are smaller than 0.05 % of the complete dataset, since such small classes are not considered to be covered representatively. Thus, we only consider those 3D points belonging to one of the six dominant semantic classes *façade*, *ground*, *cars*, *motorcycles*, *traffic signs*, and *pedestrians*, which represent a fraction of 99.81 % of the Paris-rue-Madame database (Table 6.4). A visual impression about the scene when considering these labels is provided in Fig. 6.7. Since no separation of the dataset into training set and test set is provided, we conduct a class re-balancing by randomly selecting a training set $\mathcal{X}$ with 1,000 training examples per class, while the remaining data is used as test set $\mathcal{Y}$.

Table 6.4 Number of 3D points in the most dominant classes of the Paris-rue-Madame database: both training set $\mathcal{X}$ and test set $\mathcal{Y}$ have been derived by splitting the whole dataset

Class	Training set $\mathcal{X}$	Test set $\mathcal{Y}$	Σ_{class}
Façade	1,000	9,977,435	9,978,435
Ground	1,000	8,023,295	8,024,295
Cars	1,000	1,834,383	1,835,383
Motorcycles	1,000	97,867	98,867
Traffic signs	1,000	14,480	15,480
Pedestrians	1,000	9,048	10,048
Σ_{set}	6,000	19,956,508	19,962,508

Fig. 6.7   3D point cloud of the Paris-rue-Madame database where the color encoding addresses the considered semantic classes (*façade*: *gray*; *ground*: *brown*; *cars*: *blue*; *motorcycles*: *green*; *traffic signs*: *red*; *pedestrians*: *pink*). The 3D points represented in *cyan* are those 3D points which are not considered as the respective classes are not covered representatively

6.6.1.3   Paris-Rue-Cassette Database

The *Paris-rue-Cassette database*[5] [98] represents a 3D point cloud dataset which has been acquired in January 2013 in the city of Paris, France. For data acquisition, the mobile laser scanning system Stereopolis II [97] was used which captures information about the local 3D geometry with two plane sweep LiDAR sensors (Riegl LMS-Q120i) placed on each side of the vehicle in order to mainly observe the building façades with a centimeter accuracy and a 3D LiDAR sensor (Velodyne HDL-64E) involved to observe the bottom part in between. The acquired dataset consists of 12M 3D points corresponding to a digitized street section with a length of about 200 m, and a respective annotation has been conducted in a manually assisted way in order to provide a ground truth with respect to point-wise labels organized in a hierarchy of semantic classes and segmented objects. This annotation is based on recovering a regular 2D topology for the 3D point cloud stream during data acquisition and an offline human interaction via a graph editing tool based on standard 2D image segmentation techniques [13, 131]. A further extension of the dataset has been released in the scope of a recent contest [98] and consists of about 100M 3D points belonging to 10 different zones.

For our experiments, we take into account that the hierarchy of semantic classes is provided in the form of a very detailed class tree which, in turn, is motivated by the fact that this dataset may be involved for a variety of applications. Depending on the application, it is possible to define those classes which are considered as relevant and a respective evaluation may be performed accordingly. Hence, we discard 3D points belonging to classes which are smaller than 0.05 % of the complete dataset, since such small classes are not considered to be representative, and we define our relevant classes similar to the classes considered for the Paris-rue-Madame database. As a result, we only consider those 3D points belonging to one of the seven dominant semantic classes *façade, ground, cars, 2 wheelers, road inventory, pedestrians*, and *vegetation*, which represent a fraction of 99.56 % of the Paris-rue-Cassette database (Table 6.5). A visual impression about the scene when considering these labels is provided in Fig. 6.8. Since no separation of the dataset into training set and test set is provided, we conduct a class re-balancing by randomly selecting a training set $\mathscr{X}$ with 1,000 training examples per class, while the remaining data is used as test set $\mathscr{Y}$.

6.6.2   Experiments

While a detrimental effect of an unbalanced class distribution on both feature selection and classification is avoided by the involved class re-balancing for the training set $\mathscr{X}$, the class distribution across the test set is still unbalanced for all the involved

---

[5]The Paris-rue-Cassette database and more details on the hierarchy of semantic classes are publicly available at http://data.ign.fr/benchmarks/UrbanAnalysis (last access: 30 May 2015).

Table 6.5 Number of 3D points in the most dominant classes of the Paris-rue-Cassette database: both training set $\mathscr{X}$ and test set $\mathscr{Y}$ have been derived by splitting the whole dataset

Class	Training set $\mathscr{X}$	Test set $\mathscr{Y}$	Σ_{class}
Façade	1,000	7,026,016	7,027,016
Ground	1,000	4,228,639	4,229,639
Cars	1,000	367,271	368,271
2 wheelers	1,000	39,331	40,331
Road inventory	1,000	45,105	46,105
Pedestrians	1,000	22,999	23,999
Vegetation	1,000	211,131	212,131
Σ_{set}	7,000	11,940,492	11,947,492

Fig. 6.8 3D point cloud of the Paris-rue-Cassette database where the color encoding addresses the considered semantic classes (*façade*: *gray*; *ground*: *brown*; *cars*: *blue*; *2 wheelers*: *yellow*; *road inventory*: *red*; *pedestrians*: *pink*; *vegetation*: *green*). The 3D points represented in *cyan* are those 3D points which are not considered as the respective classes are not covered representatively

datasets. For the Oakland 3D Point Cloud Dataset, for instance, 70.5 and 20.2 % of all 3D points belong to the classes *ground* and *vegetation*, while the class *façade* represents 8.4 % of the 3D points and the two remaining classes *pole/trunk* and *wire* only consist of 0.6 and 0.3 % of all 3D points, respectively. Accordingly, an overall accuracy of 70.5 % could be obtained if only those 3D points belonging to the class *ground* are correctly classified which would represent the example of an extreme overfitting. This clearly reveals that only a consideration of the overall accuracy is not sufficient to provide meaningful conclusions and therefore other measures are required for an adequate evaluation as well.

For this reason, we consider five commonly used measures for evaluation which focus on class-wise and global characteristics: (i) *recall* which represents a measure of completeness or quantity, (ii) *precision* which represents a measure of exactness or quality, (iii) F_1-*score* which combines recall and precision with equal weights, (iv) *overall accuracy* which indicates the overall performance of the respective classifier on the test set, and (v) *mean class recall* which indicates the capability of the respective classifier to detect instances of different classes. Furthermore, we may also conduct a visual inspection of the derived results. Since the classification results may slightly vary for different runs due to random sampling processes, all results are averaged over 20 runs in order to facilitate an objective comparison to other approaches. Furthermore, we take into account that, when involving filter-based feature selection, the derived feature subsets may slightly vary due to the random selection of training data in each run, and we therefore determine them as the most often occurring feature subsets over 20 runs.

Based on these considerations, we conduct our experiments. Thereby, we first consider the general behavior of the proposed method for selecting individual neighborhoods of optimal size in Sect. 6.6.3.1. Subsequently, in Sect. 6.6.3.2, we focus on the impact of 7 different neighborhood definitions given by

- the neighborhood $\mathcal{N}_{10}$ formed by the 10 closest neighbors,
- the neighborhood $\mathcal{N}_{25}$ formed by the 25 closest neighbors,
- the neighborhood $\mathcal{N}_{50}$ formed by the 50 closest neighbors,
- the neighborhood $\mathcal{N}_{75}$ formed by the 75 closest neighbors,
- the neighborhood $\mathcal{N}_{100}$ formed by the 100 closest neighbors,
- the optimal neighborhood $\mathcal{N}_{\text{opt,dim}}$ for each individual 3D point when considering dimensionality-based scale selection [28], and
- the optimal neighborhood $\mathcal{N}_{\text{opt},\lambda}$ for each individual 3D point when considering our proposed approach of eigenentropy-based scale selection

on the classification results of 10 standard classifiers belonging to different categories. In this regard, the latter two neighborhood definitions involving individual neighborhoods of optimal size are based on varying the scale parameter k between $k_{\min} = 10$ and $k_{\max} = 100$ with a step size of $\Delta k = 1$, and selecting the respective value k corresponding to the minimum Shannon entropy of the respective criterion. Since training a classifier strongly depends on the given training data, we furthermore consider the influence of a varying amount of training data on the classification results in Sect. 6.6.3.3.

In the next step, in Sect. 6.6.3.4, we focus on testing seven different feature sets for each neighborhood definition:

- the complete feature set $\mathcal{S}_{\text{all}}$ consisting of all 21 features,
- the feature subset $\mathcal{S}_{\text{dim}}$ covering the three dimensionality features of linearity L_λ, planarity P_λ and scattering S_λ,

- the feature subset $\mathscr{S}_{\lambda,3D}$ covering the eight eigenvalue-based 3D features,
- the feature subset $\mathscr{S}_5$ consisting of the five best-ranked features according to our general relevance metric presented in Sect. 6.3.3,
- the feature subset $\mathscr{S}_{CFS}$ derived via Correlation-based Feature Selection [52],
- the feature subset $\mathscr{S}_{FCBF}$ derived via the Fast Correlation-Based Filter [151], and
- the feature subset $\mathscr{S}_{mRMR}$ derived via the minimal-redundancy-maximal-relevance (mRMR) criterion [101].

The latter four feature subsets are based on either explicitly or implicitly assessing feature relevance. Note that—when involving such an approach for feature selection—the complete feature set only has to be calculated and stored for the training data, whereas a significantly smaller subset of relevant features automatically selected by the respective approach has to be calculated and stored for the test data.

This is followed by a transfer of the derived feature selection results to other datasets in Sect. 6.6.3.5. Finally, we will make considerations with respect to large-scale capability in Sect. 6.6.3.6 and also present results for an approach relying on contextual classification in Sect. 6.6.3.7.

All implementation and processing was done in Matlab. In the following, the main focus is put on the impact of both optimal neighborhood size selection and feature selection on the classification results. We may expect that (i) optimal neighborhoods for individual 3D points significantly improve the classification results and (ii) feature subsets selected via the presented feature selection approaches provide an increase in classification accuracy.

6.6.3   Results

In the following, we present the results derived by applying our main framework and the respective extensions. Thereby, we take into account a variety of different aspects and hence split the presentation of experimental results to respective subsections.

6.6.3.1   Insights Concerning Neighborhood Selection

First, we intend to provide more insights in the process of optimal neighborhood size selection. For this purpose, we involve the Oakland 3D Point Cloud Dataset [85, 86] in respective experiments. Since our approach for selecting an optimal scale parameter k involves an upper boundary of $k_{max} = 100$ in order to limit the computational costs, we might expect that it is likely to have a certain amount of 3D points favoring a higher value of k. Consequently, we consider the distribution of k across the full dataset over the specified interval between $k_{min} = 10$ and $k_{max} = 100$ with $\Delta k = 1$. The respective distribution for $\mathscr{N}_{opt,\lambda}$ is shown in Fig. 6.9 and quite similar for $\mathscr{N}_{opt,dim}$. This figure reveals a clear trend toward smaller neighborhoods, and the percentage of 3D points

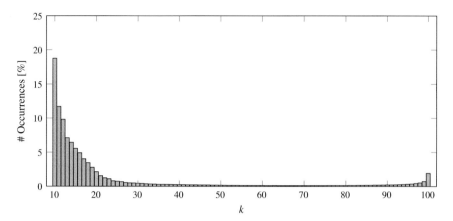

Fig. 6.9 Distribution of the assigned optimal neighborhood size k for all 3D points of the Oakland 3D Point Cloud Dataset

which are assigned neighborhoods with $k < 100$ neighbors is 98.08 and 98.13 % for $\mathscr{N}_{opt,\lambda}$ and $\mathscr{N}_{opt,dim}$. For the last bin corresponding to $k_{max} = 100$, a slight increase may be observed. The distributions per class are provided in Fig. 6.10 for $\mathscr{N}_{opt,\lambda}$ and follow the major trend with only a slight difference between different classes.

However, it has to be taken into account that—when considering optimal neighborhoods $\mathscr{N}_{opt,\lambda}$ and $\mathscr{N}_{opt,dim}$—an additional processing time of approximately 21 s and 758 s is required on a high-performance computer (Intel Core i7-3820, 3.6 GHz, 64 GB RAM) for the full training set and the test set, respectively. The additional effort is significant in comparison to feature extraction, where approximately 4 s and 2,793 s are required for calculating all features for the training set and the test set. This raises the question if individual neighborhoods of optimal size are really necessary. We focus on this issue in the next subsection.

6.6.3.2   Impact of Optimal Neighborhood Selection

Keeping in mind that involving individual neighborhoods of optimal size increases the computational burden, it is desirable to obtain an objective impression if this is really necessary. For this reason, we now focus on reasoning about the impact of neighborhood selection on the individual classification of 3D points based on the respective feature vectors. Thereby, we derive all 21 geometric features for each of the 7 different neighborhood definitions and provide the respective feature sets to 10 classifiers of different categories. The obtained overall accuracy and mean class recall values are provided in Tables 6.6 and 6.7. For an in-depth analysis concerning the impact of neighborhood selection on the classification results, the recall and precision values for the different neighborhood definitions and different classifiers are provided in Tables 6.8 and 6.9. The corresponding F_1-scores are visualized in

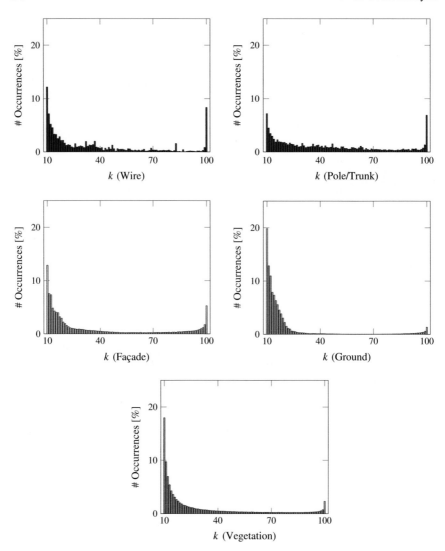

Fig. 6.10 Distribution of the assigned optimal neighborhood size k for the different classes of the Oakland 3D Point Cloud Dataset

Fig. 6.11. In order to argue about the efficiency of the involved classifiers, we also provide the absolute and relative processing times for training phase and testing phase in Table 6.10.

Note that the combination of optimal neighborhoods $\mathcal{N}_{\text{opt},\lambda}$ and a Random Forest classifier provides a good solution with respect to both accuracy and computational

Table 6.6 Overall accuracy (in %) for different neighborhood definitions and different classifiers

$\mathcal{N}$	NN	DT	NB	LDA	QDA	SVM	RF	RFe	AB	MLP
$\mathcal{N}_{10}$	73.86	65.64	78.88	87.38	78.93	82.93	87.53	81.94	86.78	80.54
$\mathcal{N}_{25}$	86.25	69.30	83.64	90.08	83.62	88.88	90.50	88.77	89.99	78.59
$\mathcal{N}_{50}$	88.89	75.47	85.03	92.83	84.95	**92.00**	91.54	90.42	91.80	85.68
$\mathcal{N}_{75}$	**89.97**	76.87	85.00	**93.05**	84.99	91.99	91.06	**91.16**	90.56	87.07
$\mathcal{N}_{100}$	89.90	**84.45**	84.33	92.60	84.43	91.76	90.16	90.59	87.01	84.39
$\mathcal{N}_{opt,dim}$	79.34	70.71	83.75	91.01	83.80	90.15	91.89	90.12	91.62	85.69
$\mathcal{N}_{opt,\lambda}$	79.87	75.76	**85.63**	90.39	**85.69**	89.10	**92.25**	90.45	**92.28**	**87.29**

Bold values indicate the highest overall accuracy obtained with the respective classifier

Table 6.7 Mean class recall values (in %) for different neighborhood definitions and different classifiers

$\mathcal{N}$	NN	DT	NB	LDA	QDA	SVM	RF	RFe	AB	MLP
$\mathcal{N}_{10}$	63.40	54.19	62.29	70.68	62.33	58.86	70.78	63.75	67.52	64.20
$\mathcal{N}_{25}$	70.01	57.41	68.46	75.54	68.47	68.50	74.65	71.48	68.64	68.03
$\mathcal{N}_{50}$	69.47	59.99	67.12	72.76	66.98	68.47	72.72	69.22	71.46	69.13
$\mathcal{N}_{75}$	68.29	57.82	65.49	73.05	65.44	68.00	69.99	68.88	68.19	70.47
$\mathcal{N}_{100}$	66.66	57.96	63.44	72.35	63.46	64.76	68.51	67.16	59.58	68.98
$\mathcal{N}_{opt,dim}$	**74.17**	62.15	74.49	81.36	74.35	79.58	81.70	78.35	77.63	78.61
$\mathcal{N}_{opt,\lambda}$	73.98	**66.99**	**76.19**	**82.05**	**76.15**	**79.97**	**82.59**	**78.70**	**79.49**	**79.92**

Bold values indicate the highest mean class recall value obtained with the respective classifier

efficiency. A visualization of respective classification results is provided in Fig. 6.12 for different parts of the scene.

6.6.3.3 Impact of Training Data

While the Random Forest classifier provides a good trade-off with respect to both accuracy and computational efficiency, we might also have to take into account if our training data obtained after the class re-balancing by randomly sampling the same number of 1,000 training examples per class is still representative. For this purpose, we consider the influence of having more or less training examples per class which, in turn, is of particular interest when dealing with datasets where small classes are only represented by a few hundreds of 3D points. Hence, we exploit such a Random Forest classifier and consider varying numbers of training examples per class. The respective recall and precision values are provided in Tables 6.11 and 6.12. Note that, for less training examples per class, the recall values tend to decrease for the smaller classes of *wire* and *pole/trunk* while the respective precision values tend to increase. The same characteristic holds when involving all training examples per class, but with a more significant impact.

Table 6.8 Recall values (in %) for different neighborhood definitions and different classifiers

Class	$\mathcal{N}$	NN	DT	NB	LDA	QDA	SVM	RF	RFe	AB	MLP
Wire	$\mathcal{N}_{10}$	65.51	55.20	65.92	77.06	65.69	62.21	70.46	72.24	58.66	65.61
	$\mathcal{N}_{25}$	64.63	50.50	70.78	73.30	70.97	60.11	69.48	68.14	52.24	63.70
	$\mathcal{N}_{50}$	50.40	42.09	66.88	48.57	66.69	41.13	56.86	55.82	53.55	51.81
	$\mathcal{N}_{75}$	49.68	35.22	63.10	48.48	63.29	43.06	49.71	51.72	51.94	48.84
	$\mathcal{N}_{100}$	50.09	26.62	60.54	46.74	60.35	34.91	49.67	52.51	54.69	47.41
	$\mathcal{N}_{opt,dim}$	77.91	69.30	75.72	82.51	75.24	80.19	85.16	76.58	78.58	78.54
	$\mathcal{N}_{opt,\lambda}$	**80.13**	**73.48**	**79.15**	**87.00**	**79.13**	**82.99**	**86.05**	**82.01**	**80.40**	**81.11**
Pole/trunk	$\mathcal{N}_{10}$	61.74	60.45	52.05	61.55	52.29	35.88	68.49	47.56	66.92	58.24
	$\mathcal{N}_{25}$	66.72	63.62	57.98	69.92	58.21	55.67	69.59	63.79	58.97	71.77
	$\mathcal{N}_{50}$	61.11	60.23	50.70	60.56	50.62	55.32	62.64	56.39	53.76	67.73
	$\mathcal{N}_{75}$	55.96	52.98	45.81	59.80	45.95	50.83	58.63	50.82	46.89	69.23
	$\mathcal{N}_{100}$	47.35	49.51	41.76	60.62	41.60	41.49	58.27	45.24	38.91	69.60
	$\mathcal{N}_{opt,dim}$	73.91	70.57	75.22	78.81	75.11	**78.42**	78.90	**75.41**	65.90	81.61
	$\mathcal{N}_{opt,\lambda}$	**74.49**	**71.36**	**76.33**	**79.85**	**76.28**	78.10	**79.99**	73.25	**70.17**	**82.07**
Façade	$\mathcal{N}_{10}$	46.16	29.42	43.35	48.21	43.45	43.45	50.29	41.94	48.62	42.42
	$\mathcal{N}_{25}$	52.10	39.51	46.70	59.97	46.49	54.15	60.98	52.35	55.89	49.72
	$\mathcal{N}_{50}$	65.27	51.29	47.32	73.81	47.29	65.00	**68.13**	55.96	**74.19**	58.94
	$\mathcal{N}_{75}$	65.19	51.16	48.47	**75.85**	48.36	65.22	67.51	63.75	70.07	**66.78**
	$\mathcal{N}_{100}$	**65.93**	**52.90**	47.06	73.97	47.28	**66.14**	62.69	61.09	34.41	65.56
	$\mathcal{N}_{opt,dim}$	61.65	33.57	**54.37**	68.82	**54.34**	64.48	65.90	**65.77**	63.91	65.75
	$\mathcal{N}_{opt,\lambda}$	55.82	46.63	51.88	67.12	52.15	64.39	67.01	58.87	65.73	66.43
Ground	$\mathcal{N}_{10}$	80.32	73.96	88.26	97.55	88.30	95.44	98.23	91.84	97.61	89.79
	$\mathcal{N}_{25}$	94.66	76.50	90.34	97.69	90.42	97.31	**98.91**	97.02	97.81	84.45
	$\mathcal{N}_{50}$	96.25	81.71	91.58	**98.55**	91.56	**98.47**	98.84	97.93	**98.52**	92.23
	$\mathcal{N}_{75}$	98.48	83.27	**91.66**	**98.55**	**91.83**	**98.47**	98.81	**98.12**	98.35	**93.55**
	$\mathcal{N}_{100}$	**98.58**	**93.72**	91.51	98.25	91.59	97.86	98.71	97.97	98.19	90.49
	$\mathcal{N}_{opt,dim}$	82.88	78.55	88.95	97.04	89.07	96.83	98.52	96.93	98.00	90.80
	$\mathcal{N}_{opt,\lambda}$	84.06	83.90	90.47	96.24	90.70	94.92	98.48	96.58	98.41	92.70
Vegetation	$\mathcal{N}_{10}$	63.27	51.90	61.86	69.03	61.91	57.31	66.45	65.15	65.79	64.94
	$\mathcal{N}_{25}$	71.95	56.95	76.49	76.83	76.24	75.27	74.29	76.11	78.28	70.53
	$\mathcal{N}_{50}$	74.32	64.62	79.09	82.33	78.76	82.44	77.10	80.00	77.29	74.92
	$\mathcal{N}_{75}$	72.13	66.48	78.41	**82.59**	77.76	82.43	75.31	79.96	73.70	73.92
	$\mathcal{N}_{100}$	71.36	**67.06**	76.32	82.19	76.48	**83.38**	73.20	78.96	71.69	71.87
	$\mathcal{N}_{opt,dim}$	74.51	58.76	78.17	79.62	78.00	77.99	79.99	77.05	81.78	76.34
	$\mathcal{N}_{opt,\lambda}$	**75.38**	59.56	**83.09**	80.02	**82.49**	79.46	**81.41**	**82.79**	**82.74**	**77.26**

Bold values indicate the highest recall value obtained with the respective classifier for the respective class

Table 6.9 Precision values (in %) for different neighborhood definitions and different classifiers

Class	$\mathcal{N}$	NN	DT	NB	LDA	QDA	SVM	RF	RFe	AB	MLP
Wire	$\mathcal{N}_{10}$	1.61	1.00	3.21	6.14	3.24	5.15	5.51	4.15	4.89	2.59
	$\mathcal{N}_{25}$	4.28	1.05	3.83	5.93	3.87	4.54	7.12	5.65	6.62	4.89
	$\mathcal{N}_{50}$	3.63	1.34	3.50	5.17	3.50	5.06	4.81	4.91	6.16	4.79
	$\mathcal{N}_{75}$	5.19	1.47	3.34	5.18	3.35	5.78	4.00	4.88	4.25	5.29
	$\mathcal{N}_{100}$	**5.29**	**2.72**	3.26	4.95	3.24	**5.79**	3.98	5.15	4.13	**5.49**
	$\mathcal{N}_{opt,dim}$	1.85	1.70	5.02	**6.27**	4.93	5.63	7.98	5.90	8.09	4.09
	$\mathcal{N}_{opt,\lambda}$	2.97	1.72	**6.49**	5.76	**6.46**	5.60	**9.03**	**7.86**	**9.34**	5.01
Pole/trunk	$\mathcal{N}_{10}$	4.48	5.18	3.75	8.10	3.67	2.43	7.99	3.00	6.60	5.51
	$\mathcal{N}_{25}$	6.28	6.38	5.78	9.32	5.68	5.99	9.46	6.24	6.36	4.77
	$\mathcal{N}_{50}$	8.66	8.46	7.97	16.61	7.90	9.02	19.47	7.53	10.02	6.71
	$\mathcal{N}_{75}$	8.17	11.09	7.91	19.26	7.89	8.85	18.25	8.84	11.10	8.14
	$\mathcal{N}_{100}$	7.19	9.04	8.27	18.55	7.97	8.17	13.55	7.42	5.06	7.50
	$\mathcal{N}_{opt,dim}$	**9.97**	6.85	**18.90**	**34.65**	**18.88**	**12.91**	22.09	**11.58**	14.86	13.55
	$\mathcal{N}_{opt,\lambda}$	9.13	**11.62**	18.22	34.52	18.38	11.46	**24.13**	11.10	**18.71**	**14.54**
Façade	$\mathcal{N}_{10}$	63.54	47.59	60.86	66.03	61.41	52.59	77.62	62.50	72.56	60.01
	$\mathcal{N}_{25}$	73.20	49.82	79.54	78.49	79.22	79.85	83.88	79.40	80.79	54.33
	$\mathcal{N}_{50}$	**78.61**	64.97	**82.78**	83.42	**82.83**	**83.16**	83.43	**81.41**	**89.05**	74.60
	$\mathcal{N}_{75}$	77.38	65.10	78.84	**87.48**	77.71	81.60	80.28	80.22	85.71	**76.78**
	$\mathcal{N}_{100}$	76.56	**68.74**	65.27	84.63	67.69	72.93	76.19	74.78	71.15	68.03
	$\mathcal{N}_{opt,dim}$	73.45	45.95	79.62	79.73	79.63	77.27	83.71	77.04	82.92	74.88
	$\mathcal{N}_{opt,\lambda}$	71.56	51.84	82.71	80.75	81.80	76.14	**84.69**	76.69	84.21	76.42
Ground	$\mathcal{N}_{10}$	98.76	98.40	96.68	96.78	96.73	98.68	96.82	99.64	97.57	96.47
	$\mathcal{N}_{25}$	99.01	98.24	**99.67**	99.58	**99.67**	99.68	**98.58**	99.75	**99.16**	98.57
	$\mathcal{N}_{50}$	98.84	98.23	98.27	99.51	97.98	99.75	97.77	99.74	98.66	98.83
	$\mathcal{N}_{75}$	98.66	97.70	98.63	99.42	98.27	99.73	97.86	99.66	97.67	**99.28**
	$\mathcal{N}_{100}$	98.56	96.67	97.84	**99.64**	97.93	**99.82**	97.92	99.74	96.45	99.19
	$\mathcal{N}_{opt,dim}$	**99.16**	**98.78**	96.58	96.62	96.60	99.39	97.67	**99.76**	98.15	97.97
	$\mathcal{N}_{opt,\lambda}$	99.04	98.45	96.72	96.25	96.60	99.22	97.18	99.60	97.57	97.57
Vegetation	$\mathcal{N}_{10}$	77.23	76.16	**81.06**	95.34	**81.59**	92.46	94.79	91.73	93.32	93.84
	$\mathcal{N}_{25}$	91.55	77.87	78.06	94.89	78.20	94.47	94.87	**94.79**	94.29	90.67
	$\mathcal{N}_{50}$	**93.06**	82.52	79.23	93.16	79.71	93.47	94.40	94.18	93.36	90.37
	$\mathcal{N}_{75}$	92.93	80.16	77.69	91.32	78.79	92.08	93.84	93.30	93.34	86.55
	$\mathcal{N}_{100}$	92.58	79.16	78.91	89.95	78.69	91.14	93.55	92.88	92.82	81.99
	$\mathcal{N}_{opt,dim}$	87.28	78.70	69.87	95.87	70.11	**95.12**	94.97	94.97	93.75	94.65	93.06
	$\mathcal{N}_{opt,\lambda}$	75.77	**84.53**	73.77	**96.27**	74.32	93.96	**95.87**	92.83	**95.19**	**93.96**

Bold values indicate the highest precision value obtained with the respective classifier for the respective class

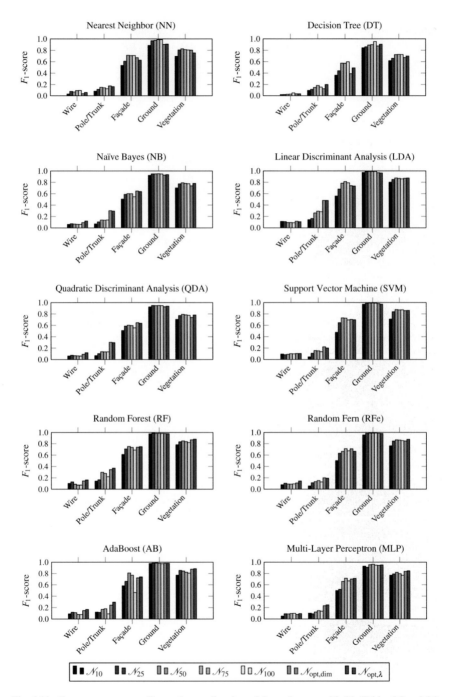

Fig. 6.11 F_1-scores corresponding to the recall and precision values provided in Tables 6.8 and 6.9

Table 6.10 Absolute and relative processing times for training phase and testing phase when using different classifiers on a standard notebook (Intel Core i5-2410M, 2.3 GHz, 4 GB RAM)

Time	NN	DT	NB	LDA	QDA	SVM	RF	RFe	AB	MLP
t_{train} (s)	00.00	0.11	0.01	0.05	0.07	1.39	0.44	0.03	6.20	2.28
t_{train} (%)	00.00	24.54	3.13	10.74	15.18	317.19	100.00	6.96	1410.66	518.13
t_{test} (s)	167.52	0.65	3.71	4.45	3.92	319.48	6.33	8.12	76.31	1.80
t_{test} (%)	2645.77	10.22	58.64	70.34	61.96	5045.68	100.00	128.22	1205.24	28.45

The reference for relative values is represented by the Random Forest classifier. Note that, for the training, additional time is required in order to tune the settings of some classifiers (SVM, RF, RFe, AB and MLP)

Fig. 6.12 Exemplary classification results when using optimal neighborhoods $\mathcal{N}_{opt,\lambda}$ and a Random Forest classifier (*wire*: *blue*; *pole/trunk*: *red*; *façade*: *gray*; *ground*: *brown*; *vegetation*: *green*)

Table 6.11 Recall values (in %) for eigenentropy-based scale selection combined with a Random Forest classifier

Class	250 samples	500 samples	750 samples	1,000 samples	All samples
Wire	84.22	85.07	85.91	86.05	82.31
Pole/trunk	76.47	77.98	78.77	79.99	70.52
Façade	66.50	67.14	67.47	67.01	65.80
Ground	98.50	98.55	98.49	98.48	97.89
Vegetation	81.82	81.34	81.78	81.40	93.35

Table 6.12 Precision values (in %) for eigenentropy-based scale selection combined with a Random Forest classifier

Class	250 samples	500 samples	750 samples	1,000 samples	All samples
Wire	9.06	9.21	9.05	9.03	10.19
Pole/trunk	32.38	28.53	27.83	24.14	45.75
Façade	82.22	82.52	84.32	84.69	92.73
Ground	96.77	96.94	97.01	97.18	98.98
Vegetation	95.48	95.72	95.73	95.87	89.82

6.6.3.4   Impact of Feature Selection

Focusing on the capability of our framework toward large-scale 3D scene analysis, where significantly larger datasets may be expected, we also intend to provide a solution for selecting relevant features and discarding irrelevant ones in order to reduce the computational burden with respect to processing time and memory consumption. For this purpose, we first consider our proposed general relevance metric in combination with features derived via the neighborhood definition $\mathcal{N}_{50}$ relying on the 50 closest neighbors. The respective ranking of the features according to their suitability is depicted in Fig. 6.13. For a respective 3D scene analysis, we follow the principle of Sequential Forward Selection (SFS), i.e., we begin with a feature set consisting of only the most relevant feature, train a respective classifier and test the performance of the classifier on the test set. Subsequently, the second best-ranked feature is added to the involved feature set, a new classifier is trained and the performance of the new classifier is evaluated on the test set. This procedure is repeated until the feature set contains all 21 defined low-level geometric 3D and 2D features. Between the features with global rank 5 and global rank 6, the relevance metric shows a significant change (Fig. 6.13). For this reason, we assume the five best-ranked features to be a meaningful subset in the experiments. Exemplary classification results obtained for the different feature subsets when using different classifiers in the form of a Nearest Neighbor (NN) classifier, a k Nearest Neighbor (k-NN) classifier with $k = 50$, a Naïve Bayesian (NB) classifier, and a Support Vector Machine (SVM) classifier are depicted in Fig. 6.14.

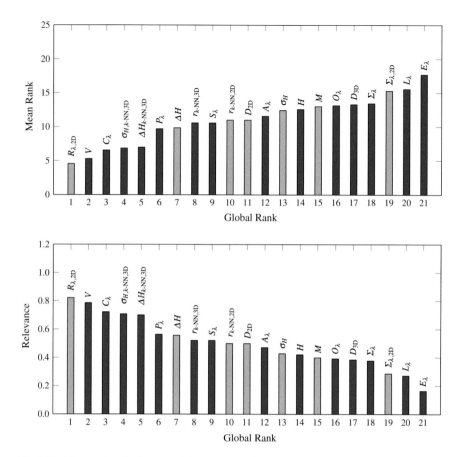

Fig. 6.13 Mean rank and relevance of the derived 3D features (*blue*) and those features derived via considerations in 2D (*green*)

Based on these simple tests focusing on our general relevance metric, we may consider a more detailed evaluation addressing different feature selection approaches, where we also take into account the results of previous experiments. Accordingly, we again use a Random Forest classifier and provide the respective results obtained when involving the 7 different neighborhood definitions and 7 different feature sets. This yields a total number of 49 possible combinations. For each combination, the respective overall accuracy and mean class recall values are provided in Tables 6.13 and 6.14. Note that for this experiment

- $\mathcal{S}_{\text{all}}$ contains all 21 features,
- $\mathcal{S}_{\text{dim}}$ contains 3 features (which represents about 14 % of the available features),
- $\mathcal{S}_{\lambda,\text{3D}}$ contains 8 features (about 38 %),
- $\mathcal{S}_{5}$ contains 5 features (about 24 %),
- $\mathcal{S}_{\text{CFS}}$ contains between 12 and 16 features (about 57–76 %),
- $\mathcal{S}_{\text{FCBF}}$ contains between 6 and 9 features (about 29–43 %), and

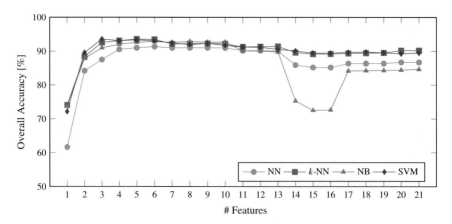

Fig. 6.14 Overall classification accuracy for different feature subsets and different classifiers: the size of the feature subset is iteratively increased according to the ascending global rank depicted in Fig. 6.13

Table 6.13 Overall accuracy (in %) when using a Random Forest classifier, different neighborhood definitions and different feature sets

$\mathcal{N}$	$\mathcal{S}_{all}$	$\mathcal{S}_{dim}$	$\mathcal{S}_{\lambda,3D}$	$\mathcal{S}_5$	$\mathcal{S}_{CFS}$	$\mathcal{S}_{FCBF}$	$\mathcal{S}_{mRMR}$
$\mathcal{N}_{10}$	87.50	58.36	74.33	85.66	87.43	87.29	82.11
$\mathcal{N}_{25}$	90.78	68.80	82.48	89.60	90.59	91.78	84.70
$\mathcal{N}_{50}$	91.64	73.19	81.38	91.01	91.71	92.69	85.64
$\mathcal{N}_{75}$	91.00	**73.63**	80.12	90.24	91.17	91.47	85.99
$\mathcal{N}_{100}$	90.11	72.99	81.96	89.84	90.31	90.94	85.76
$\mathcal{N}_{opt,dim}$	91.92	69.59	77.69	91.41	91.83	91.55	**86.82**
$\mathcal{N}_{opt,\lambda}$	**92.28**	63.61	**84.88**	**91.44**	**92.27**	**92.78**	84.28

Bold values indicate the highest overall accuracy obtained for the respective feature set

- $\mathcal{S}_{mRMR}$ contains 10 features (about 48 %).

In this regard, we want to point out that the latter four subsets contain features which are distributed across both 3D and 2D features.

6.6.3.5   Transfer Between Datasets

Based on our observations in the previous subsections, we may also focus on the transfer of the derived feature selection results in order to classify a different 3D point cloud. Accordingly, we involve the Paris-rue-Madame database [121] and we again select a Random Forest classifier as it represents an efficient classifier with respect to both accuracy and computational effort. Since in all previous experiments, our approach for selecting individual neighborhoods of optimal size ($\mathcal{N}_{opt,\lambda}$) has proven to have a beneficial impact on the respective results, we select this neighborhood

Table 6.14 Mean class recall values (in %) when using a Random Forest classifier, different neighborhood definitions and different feature sets

$\mathcal{N}$	$\mathcal{S}_{all}$	$\mathcal{S}_{dim}$	$\mathcal{S}_{\lambda,3D}$	$\mathcal{S}_5$	$\mathcal{S}_{CFS}$	$\mathcal{S}_{FCBF}$	$\mathcal{S}_{mRMR}$
$\mathcal{N}_{10}$	70.83	48.41	59.95	58.24	69.28	70.08	62.46
$\mathcal{N}_{25}$	75.48	55.68	65.22	73.30	74.24	76.58	63.61
$\mathcal{N}_{50}$	72.71	54.41	64.43	65.91	72.64	74.14	60.90
$\mathcal{N}_{75}$	69.75	52.12	61.37	59.86	70.19	68.66	58.35
$\mathcal{N}_{100}$	68.49	50.33	61.37	60.22	69.02	66.34	56.05
$\mathcal{N}_{opt,dim}$	81.79	**61.53**	67.65	75.57	81.33	80.83	**74.72**
$\mathcal{N}_{opt,\lambda}$	**82.60**	59.48	**69.17**	**78.50**	**82.39**	**82.93**	69.37

Bold values indicate the highest mean class recall value obtained for the respective feature set

definition in order to extract the geometric 3D and 2D features. For the same training data and test data, we compare the results obtained for the full feature set $\mathcal{S}_{all}$ and the most powerful feature selection approaches yielding the feature sets $\mathcal{S}_{CFS}$ and $\mathcal{S}_{FCBF}$ in Table 6.15. Corresponding to the provided recall and precision values, we obtain

- an overall accuracy of 88.76 % and a mean class recall of 83.56 % for the full feature set $\mathcal{S}_{all}$,
- an overall accuracy of 88.98 % and a mean class recall of 84.66 % for the feature set $\mathcal{S}_{CFS}$, and
- an overall accuracy of 89.16 % and a mean class recall of 83.83 % for the feature set $\mathcal{S}_{FCBF}$.

In order to obtain an impression about the quality of the derived results, the results when involving the full feature set $\mathcal{S}_{all}$ are visualized in Fig. 6.15.

6.6.3.6   Transfer to Large-Scale 3D Scene Analysis

When involving larger datasets such as the Paris-rue-Madame database [121] or the Paris-rue-Cassette database [98], we may also be interested in obtaining an impres-

Table 6.15 Recall values R and precision values P (in %) when using eigenentropy-based scale selection, different feature sets and a Random Forest classifier

Class	R_{all}	P_{all}	R_{CFS}	P_{CFS}	R_{FCBF}	P_{FCBF}
Façade	95.29	96.45	95.11	96.61	95.23	96.89
Ground	86.69	97.92	86.38	98.70	87.44	98.66
Cars	63.13	78.96	67.71	77.42	64.42	79.82
Motorcycles	71.52	9.25	73.68	9.51	73.30	8.45
Traffic signs	95.82	4.76	95.88	4.90	96.26	5.34
Pedestrians	88.89	1.67	89.21	1.74	86.32	1.75

Fig. 6.15 Exemplary classification results when using optimal neighborhoods $\mathcal{N}_{\mathrm{opt},\lambda}$, the full feature set $\mathcal{S}_{\mathrm{all}}$ and a Random Forest classifier (*façade*: *gray*; *ground*: *brown*; *cars*: *blue*; *motorcycles*: *green*; *traffic signs*: *red*; *pedestrians*: *pink*)

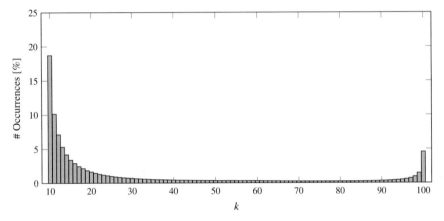

Fig. 6.16 Distribution of the assigned optimal neighborhood size k for all 3D points of the Paris-rue-Madame database

sion about the distribution of the scale parameter k across all 3D points. Accordingly, we provide a visualization of the respective distributions obtained for the Paris-rue-Madame database with 20M 3D points and the Paris-rue-Cassette database with 12M 3D points in the form of histograms in Figs. 6.16 and 6.17. These figures clearly reveal a trend toward smaller values of k. Since the last bin in the histograms ($k_{\mathrm{max}} = 100$) is likely to also represent those 3D points which might favor a higher value than $k_{\mathrm{max}} = 100$, a small increase may be observed at the upper boundary of the considered interval. However, the percentage of 3D points which are assigned an optimal neighborhood with $k < 100$ neighbors is 95.44 and 98.72 % for the Paris-rue-Madame database and the Paris-rue-Cassette database, respectively. This indicates that our selection of k_{max} may be interpreted as appropriate.

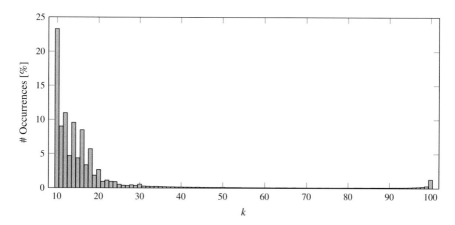

Fig. 6.17 Distribution of the assigned optimal neighborhood size k for all 3D points of the Paris-rue-Cassette database

While the provided histograms only reveal a trend toward smaller values of k, a more detailed analysis may be desirable. Consequently, we also visualize the behavior of the scale parameter k for the different classes of the Paris-rue-Cassette database in Fig. 6.18. Even though such a class-wise consideration reveals a slight difference between the different classes, there is no clear indication of a characteristic which could be regarded as specific for a certain class. Furthermore, we provide a qualitative visualization of the distributions of the assigned optimal neighborhood size across all 3D points of both involved datasets in Figs. 6.19 and 6.20. In these figures, a significantly smoother behavior may be observed for the Paris-rue-Cassette database.

With the intention of classifying such large-scale datasets, it is advisable to involve classifiers with a high predictive accuracy and thereby favor efficient classifiers. Furthermore, we explicitly intend to focus on applicability, reproducibility, and scalability of all involved methods in order to facilitate 3D scene analysis in large-scale urban environments so that nonexpert users may apply our framework as well. Consequently, the involved classification schemes should be easy to use without crucial parameter selection and respective implementations should be available in different software tools. Based on our previous experiments, we may state that (i) the use of individually optimized 3D neighborhoods derived via our proposed approach of eigenentropy-based scale selection increases the quality of the classification results the most and (ii) a Random Forest classifier satisfies the constraints with respect to accuracy and efficiency, while its simplicity allows a rather intuitive interpretation of the classification principle. For these reasons, we now focus on classifying both datasets based on the combination of eigenentropy-based scale selection, the complete feature set consisting of 21 geometric 3D and 2D features and a Random Forest classifier, where we consider the results for a single run due to an expected significant increase of the computational burden. Thereby, we achieve an overall accuracy of 88.82 % for the Paris-rue-Madame database and an overall accuracy of 89.60 % for the Paris-rue-Cassette database, respectively. The resulting recall and precision values as well as the corresponding F_1-scores are provided in Tables 6.16 and 6.17.

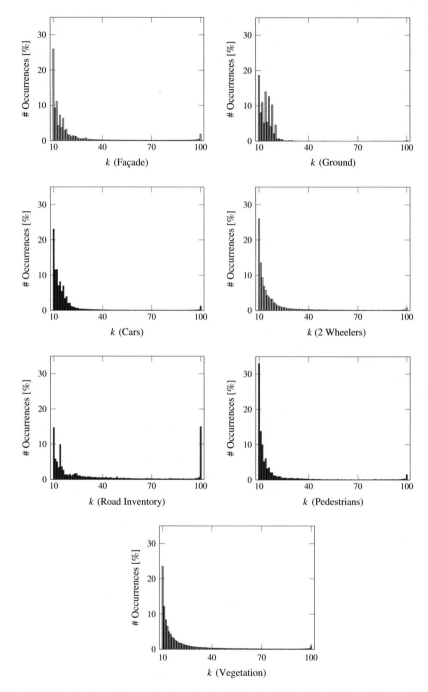

Fig. 6.18 Distribution of the assigned optimal neighborhood size k for different classes of the Paris-rue-Cassette database

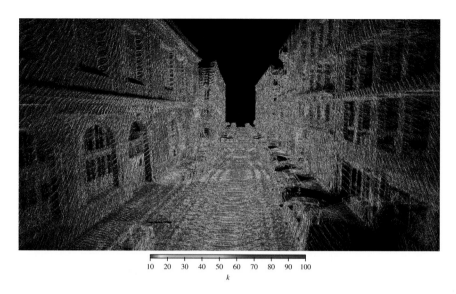

10   20   30   40   50   60   70   80   90   100
k

Fig. 6.19 Qualitative distribution of the assigned optimal neighborhood size k for all 3D points of the Paris-rue-Madame database

10   20   30   40   50   60   70   80   90   100
k

Fig. 6.20 Qualitative distribution of the assigned optimal neighborhood size k for all 3D points of the Paris-rue-Cassette database

Table 6.16 Recall values, precision values and F_1-scores (in %) for the Paris-rue-Madame database when using eigenentropy-based scale selection, the complete feature set, and a Random Forest classifier

Class	Recall	Precision	F_1-score
Façade	95.27	96.20	95.73
Ground	86.50	97.82	91.82
Cars	64.76	79.48	71.37
Motorcycles	71.98	9.80	17.25
Traffic signs	94.85	4.91	9.34
Pedestrians	87.80	1.63	3.20

Table 6.17 Recall values, precision values and F_1-scores (in %) for the Paris-rue-Cassette database when using eigenentropy-based scale selection, the complete feature set, and a Random Forest classifier

Class	Recall	Precision	F_1-score
Façade	87.21	99.28	92.85
Ground	96.46	99.24	97.83
Cars	61.12	67.67	64.23
2 wheelers	82.85	17.74	29.23
Road inventory	76.57	14.95	25.01
Pedestrians	82.25	9.24	16.61
Vegetation	86.02	25.66	39.53

Accordingly, the mean class recall value results in 83.53 % for the Paris-rue-Madame database, whereas a mean class recall value of 81.78 % is obtained for the Paris-rue-Cassette database. Finally, a visual impression about the quality of the derived results for an individual classification of 3D points based on a Random Forest classifier is depicted in Figs. 6.21 and 6.22.

The experiments have been conducted on a high-performance computer (Intel Core i7-3820, 3.6 GHz, 64 GB RAM) and, since both datasets represent almost straight street sections, we use our proposed slicing approach where the slices have a width of 10 m and the padding has a width of 0.50 m. Whereas our first prototype is based on a full and straightforward Matlab implementation, deeper investigations revealed that a significant speedup may be achieved in two ways. First, for the considered small 3D point sets formed by up to $k_{max} = 100$ neighboring 3D points, a considerable speedup in the calculation of the respective 3D covariance matrices results from simply replacing the internal Matlab function `cov` with the respective vectorized straightforward implementation. Second, an optimized approach may be based on a respective C++ implementation and, for comparison, we involve this optimization in the form of respective binaries in the test environment in Matlab. Considering both optimized approaches for the example involving the Paris-rue-Cassette database, the respectively achieved speedup clearly becomes visible in Table 6.18,

Fig. 6.21 Paris-rue-Madame database: classified point cloud with assigned semantic labels (*façade*: *gray*; *ground*: *brown*; *cars*: *blue*; *motorcycles*: *yellow*; *traffic signs*: *red*; *pedestrians*: *pink*). The 3D points represented in *cyan* are those 3D points which are not considered as the respective classes are not covered representatively. The noisy appearance results from individual classification

Fig. 6.22 Paris-rue-Cassette database: classified point cloud with assigned semantic labels (*façade*: *gray*; *ground*: *brown*; *cars*: *blue*; *2 wheelers*: *yellow*; *road inventory*: *red*; *pedestrians*: *pink*; *vegetation*: *green*). The 3D points represented in *cyan* are those 3D points which are not considered as the respective classes are not covered representatively. The noisy appearance results from individual classification

Table 6.18 Computational effort for processing the Paris-rue-Cassette database

Time	Prototype Matlab version	Optimized Matlab version	C++ implementation
t_1	27.45 h	10.90 h	2.11 h
t_2	11.84 h	10.75 h	4.28 h
t_3	~1–2 s	~1–2 s	~1–2 s
t_4	~90 s	~90 s	~90 s

The required processing times t_1 for eigenentropy-based scale selection, t_2 for feature extraction, t_3 for training on the small training set and t_4 for testing on the respective test set are listed for different implementations. Note that t_1 and t_2 correspond to a successive processing of all slices, and that t_3 and t_4 do not change since they are not affected by our optimization

where the processing times for the different subtasks are listed. Since the training phase will not change with larger datasets, the respective classification will only remain a question of computational and not human effort.

Thus, taking a tile of defined size as a reference area would even allow us to extrapolate the resulting computational effort for data processing (which also accounts for those points in the padding), and we would for instance be able to extrapolate the computational effort to full cities.

6.6.3.7   Impact of Contextual Information

Finally, we focus on the impact of optimal neighborhood size selection on the results of contextual classification. Accordingly, we involve five different neighborhood definitions in order to calculate the geometric 3D and 2D features. Three of them—denoted by $\mathcal{N}_{10}$, $\mathcal{N}_{50}$ and $\mathcal{N}_{100}$—are based on using an identical scale parameter across all considered 3D points and thus a fixed neighborhood size of $k = 10, 50$ and 100 closest neighbors. The other two neighborhood definitions are represented by $\mathcal{N}_{\text{opt,dim}}$ where the optimal neighborhood is derived via dimensionality-based scale selection and $\mathcal{N}_{\text{opt},\lambda}$ where the optimal neighborhood is derived via eigenentropy-based scale selection (see Sect. 6.3.1).

The different neighborhood definitions result in different variants of the feature vectors, and we consider the classification results for two variants of the Random Forest classifier based on different settings. The first variant—denoted as RF_{100}—represents a Random Forest classifier consisting of $N_T = 100$ decision trees with a maximum tree depth of $d_{\max} = 4$ which are trained on a training set $\mathcal{X}$ comprising 1,000 training examples per class. In contrast, the second variant—denoted as RF_{200}—represents a Random Forest classifier consisting of $N_T = 200$ decision trees with a maximum tree depth of $d_{\max} = 15$ which are trained on a training set $\mathcal{X}$ comprising 10,000 training examples per class. In both variants, a node is only split if it is reached by at least $n_{\min} = 20$ training examples, and the number n_a of features for each test is set to the square root of the number of features. As a result,

Table 6.19 Overall accuracy (in %) for different neighborhood definitions and different classifiers

$\mathscr{N}$	RF_{100}	CRF_{100}^1	CRF_{100}^5	CRF_{100}^{10}	RF_{200}	CRF_{200}^1	CRF_{200}^5	CRF_{200}^{10}
$\mathscr{N}_{10}$	85.30	89.96	89.15	90.26	88.62	90.67	91.50	91.40
$\mathscr{N}_{50}$	**91.18**	92.79	93.51	93.15	**93.85**	94.70	94.81	94.56
$\mathscr{N}_{100}$	90.04	91.15	90.78	90.73	92.47	93.26	94.39	93.24
$\mathscr{N}_{opt,dim}$	90.82	**94.50**	94.29	**95.05**	92.35	95.07	94.72	**95.25**
$\mathscr{N}_{opt,\lambda}$	91.04	94.15	**94.50**	94.49	93.48	**95.35**	**95.45**	94.88

Bold values indicate the highest overall accuracy obtained with the respective classifier

we may expect that the first variant RF_{100} represents a more efficient but less accurate variant, whereas the second variant RF_{200} probably leads to a slightly improved performance due to the larger number of considered training examples and due to the larger number of involved decision trees, though at the cost of a higher computational effort.

In a first step, we apply a classification which only relies on the association potentials and thus the unary term of the CRF, i.e., the classification based on the two variants of the Random Forest classifier which are denoted by RF_{100} and RF_{200}, respectively. Subsequently, we also consider the interaction potentials relying on the contrast-sensitive Potts model which allows a contextual 3D point cloud classification based on a CRF. In this regard, we set the second weight parameter to $w_2 = 0.5$, a value found empirically[6] which results in an equal influence of the data-dependent term and the data-independent term of the interaction potential. For the first weight parameter w_1, we consider three different values ($w_1 = 1.0$, $w_1 = 5.0$ and $w_1 = 10.0$) in order to show its impact on the classification results. The respective CRF variants are referred to as $CRF_{N_T}^1$, $CRF_{N_T}^5$ and $CRF_{N_T}^{10}$, respectively, where N_T is either 100 or 200, depending on whether the association potential relies on RF_{100} or RF_{200}. The size of the neighborhood for each node of the graph is based on the one for the definition of the features, but thresholded by a parameter $k_{max,CRF}$. For the neighborhood definition $\mathscr{N}_{10}$, we connect each point to its 10 closest neighbors, whereas for $\mathscr{N}_{50}$ and $\mathscr{N}_{100}$ the number of neighbors is set to $k_{max,CRF} = 15$ for performance reasons. For the other two neighborhood definitions $\mathscr{N}_{opt,dim}$ and $\mathscr{N}_{opt,\lambda}$, we use $k_{max,CRF} = 25$, but vary the size of the neighborhood according to the one used for the definition of the features. This results in an average number of $N_a = 21$ neighbors for $\mathscr{N}_{opt,dim}$ and $N_a = 15$ neighbors for $\mathscr{N}_{opt,\lambda}$.

As a consequence of these settings, we carry out 40 experiments. The obtained overall accuracy and mean class recall values for different neighborhood definitions and the different classifiers are provided in Tables 6.19 and 6.20, and the respective recall and precision values are provided in Tables 6.21 and 6.22. A visualization of the respective F_1-scores is provided in Fig. 6.23, and some classification results are visualized in Fig. 6.24.

---

[6]In a set of experiments not reported here, we found that changes of that parameter had very little influence on the classification results.

Table 6.20 Mean class recall values (in %) for different neighborhood definitions and different classifiers

$\mathcal{N}$	RF_{100}	CRF^1_{100}	CRF^5_{100}	CRF^{10}_{100}	RF_{200}	CRF^1_{200}	CRF^5_{200}	CRF^{10}_{200}
$\mathcal{N}_{10}$	67.05	72.35	71.69	72.75	69.27	72.52	72.20	72.54
$\mathcal{N}_{50}$	72.89	73.92	75.49	75.34	71.63	71.39	71.73	70.74
$\mathcal{N}_{100}$	67.89	68.22	67.14	67.63	65.12	64.01	67.57	64.49
$\mathcal{N}_{opt,dim}$	79.94	85.52	85.40	86.13	81.47	85.58	85.27	85.48
$\mathcal{N}_{opt,\lambda}$	**80.79**	**86.31**	**87.30**	**86.29**	**82.53**	**86.88**	**86.83**	**86.35**

Bold values indicate the highest mean class recall value obtained with the respective classifier

Table 6.21 Recall values (in %) for different neighborhood definitions and different classifiers

Class	$\mathcal{N}$	RF_{100}	CRF^1_{100}	CRF^5_{100}	CRF^{10}_{100}	RF_{200}	CRF^1_{200}	CRF^5_{200}	CRF^{10}_{200}
Wire	$\mathcal{N}_{10}$	63.15	62.89	62.31	64.26	67.95	68.19	67.21	67.40
	$\mathcal{N}_{50}$	64.44	64.07	63.78	64.42	47.73	47.34	46.39	45.76
	$\mathcal{N}_{100}$	56.72	54.85	56.88	59.04	39.98	34.79	39.17	38.43
	$\mathcal{N}_{opt,dim}$	81.29	85.00	83.53	85.64	85.19	87.51	87.19	87.11
	$\mathcal{N}_{opt,\lambda}$	**84.11**	**87.14**	**90.46**	**86.16**	**86.87**	**89.51**	**88.80**	**88.85**
Pole/trunk	$\mathcal{N}_{10}$	54.86	60.82	60.80	58.84	56.52	61.16	59.23	59.44
	$\mathcal{N}_{50}$	49.15	46.11	49.22	50.40	52.50	47.71	48.61	47.16
	$\mathcal{N}_{100}$	39.85	39.19	35.51	36.75	40.45	38.37	37.96	37.01
	$\mathcal{N}_{opt,dim}$	73.57	79.00	78.97	78.27	**73.33**	76.11	74.99	74.44
	$\mathcal{N}_{opt,\lambda}$	**74.66**	**81.65**	**82.19**	**80.79**	72.38	**78.66**	**78.76**	**77.06**
Façade	$\mathcal{N}_{10}$	53.74	57.92	60.34	60.85	38.68	43.23	43.36	44.69
	$\mathcal{N}_{50}$	**76.45**	**78.50**	**82.25**	**81.84**	**69.04**	69.77	71.35	68.38
	$\mathcal{N}_{100}$	73.66	72.58	70.10	68.08	59.81	57.33	71.21	57.83
	$\mathcal{N}_{opt,dim}$	70.14	76.20	76.81	77.31	62.96	70.96	71.47	72.19
	$\mathcal{N}_{opt,\lambda}$	70.54	76.07	75.64	76.34	65.11	**72.37**	**72.35**	**72.39**
Ground	$\mathcal{N}_{10}$	93.89	95.73	95.78	95.98	95.04	95.58	96.75	96.34
	$\mathcal{N}_{50}$	97.47	97.75	98.01	98.21	98.29	98.59	98.47	98.47
	$\mathcal{N}_{100}$	**97.91**	98.22	**98.42**	98.19	**98.66**	**99.02**	**98.95**	**99.03**
	$\mathcal{N}_{opt,dim}$	97.31	**98.43**	97.80	**98.53**	97.04	98.08	97.58	98.13
	$\mathcal{N}_{opt,\lambda}$	97.68	98.00	98.13	98.05	97.97	98.14	98.19	97.40
Vegetation	$\mathcal{N}_{10}$	69.62	84.37	79.22	83.83	88.18	94.44	94.47	94.84
	$\mathcal{N}_{50}$	76.92	83.18	84.19	81.82	**90.57**	93.53	93.85	93.91
	$\mathcal{N}_{100}$	71.32	76.25	74.80	76.11	86.71	90.52	90.55	90.16
	$\mathcal{N}_{opt,dim}$	**77.37**	**88.95**	89.91	**90.90**	88.84	95.25	95.11	95.54
	$\mathcal{N}_{opt,\lambda}$	76.95	88.68	**90.06**	90.13	90.30	**95.73**	**96.06**	**96.06**

Bold values indicate the highest recall value obtained with the respective classifier for the respective class

Table 6.22 Precision values (in %) for different neighborhood definitions and different classifiers

Class	$\mathcal{N}$	RF_{100}	CRF^1_{100}	CRF^5_{100}	CRF^{10}_{100}	RF_{200}	CRF^1_{200}	CRF^5_{200}	CRF^{10}_{200}
Wire	$\mathcal{N}_{10}$	4.44	6.32	6.02	6.48	6.94	9.65	10.69	10.16
	$\mathcal{N}_{50}$	4.77	4.61	5.24	5.00	**9.26**	**11.31**	**10.92**	**10.95**
	$\mathcal{N}_{100}$	3.73	3.90	4.14	4.31	5.98	7.09	7.58	7.01
	$\mathcal{N}_{opt,dim}$	5.18	6.78	5.91	**11.88**	7.47	10.12	8.67	10.31
	$\mathcal{N}_{opt,\lambda}$	**7.50**	**9.72**	**11.72**	10.46	8.25	9.99	10.67	10.12
Pole/trunk	$\mathcal{N}_{10}$	8.07	10.14	8.45	9.94	12.64	14.57	14.02	15.34
	$\mathcal{N}_{50}$	31.83	55.07	45.08	32.94	**41.96**	**66.19**	**75.08**	**71.52**
	$\mathcal{N}_{100}$	27.00	44.66	36.67	28.68	13.59	16.19	44.02	20.49
	$\mathcal{N}_{opt,dim}$	**35.51**	**64.99**	**47.38**	40.15	31.56	47.11	51.26	50.76
	$\mathcal{N}_{opt,\lambda}$	33.44	52.06	34.76	**44.90**	37.35	62.72	59.73	66.28
Façade	$\mathcal{N}_{10}$	57.75	82.16	77.38	79.80	74.05	83.03	82.45	83.12
	$\mathcal{N}_{50}$	**74.02**	83.56	84.40	82.37	**88.91**	91.73	91.97	92.33
	$\mathcal{N}_{100}$	66.54	71.70	68.20	69.04	83.86	88.35	89.89	88.00
	$\mathcal{N}_{opt,dim}$	67.18	**84.34**	**89.49**	**89.37**	83.80	93.35	93.03	92.76
	$\mathcal{N}_{opt,\lambda}$	69.32	82.56	87.14	88.90	87.71	**94.43**	**94.55**	**94.12**
Ground	$\mathcal{N}_{10}$	98.24	98.27	98.29	98.44	98.87	98.81	98.44	99.32
	$\mathcal{N}_{50}$	99.70	99.50	**99.94**	**99.92**	98.99	98.56	98.93	98.72
	$\mathcal{N}_{100}$	99.14	98.64	98.34	98.37	98.81	97.77	98.86	99.23
	$\mathcal{N}_{opt,dim}$	**99.77**	**99.91**	99.85	98.62	**99.77**	**99.92**	**99.90**	99.89
	$\mathcal{N}_{opt,\lambda}$	97.78	98.43	98.43	98.15	99.52	99.89	**99.90**	**99.90**
Vegetation	$\mathcal{N}_{10}$	84.51	89.96	91.64	91.76	79.71	83.39	86.98	83.79
	$\mathcal{N}_{50}$	88.36	91.57	91.84	93.18	86.16	88.10	87.39	86.82
	$\mathcal{N}_{100}$	91.06	92.05	92.84	92.44	87.99	90.06	88.09	84.90
	$\mathcal{N}_{opt,dim}$	93.52	96.50	**97.25**	**96.39**	87.78	91.84	91.87	**92.35**
	$\mathcal{N}_{opt,\lambda}$	**94.44**	**96.60**	96.89	96.30	**90.45**	92.24	92.01	89.92

Bold values indicate the highest precision value obtained with the respective classifier for the respective class

6.7 Discussion

Since our experiments address many different aspects, we split our discussion of the derived results into respective subsections addressing the main framework (Sect. 6.7.1), the extension toward data-intensive processing (Sect. 6.7.2) and the use of contextual information (Sect. 6.7.3).

6.7.1 Our Main Framework

First, we discuss the results derived with our main framework designed for a semantic 3D point cloud interpretation based on optimal neighborhoods, relevant features

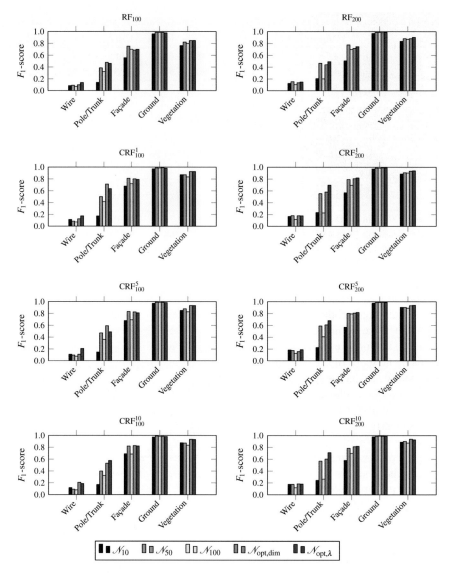

Fig. 6.23 F_1-scores corresponding to the recall and precision values provided in Tables 6.21 and 6.22

and efficient classifiers [142]. Certainly, a huge advantage of our main framework focusing on the individual classification of 3D points only based on the respective feature vectors is given by its composition of four successive components, where each component may be treated independently from the others which, in turn, allows to exhaustively test all conceivable configurations. For each of these components, we briefly discuss the main conclusions derived from our experiments.

Fig. 6.24 Classified 3D point clouds for the neighborhoods $\{\mathcal{N}_{50}, \mathcal{N}_{opt,\lambda}\}$ (*left* and *right* column) and the classifiers $\{RF_{200}, CRF_{200}^5\}$ (*top* and *bottom* row) when using our standard color encoding (*wire*: *blue*; *pole/trunk*: *red*; *façade*: *gray*; *ground*: *brown*; *vegetation*: *green*). Note the noisy appearance of the results for individual point classification (*top* row)

For the first component of neighborhood selection, we may conclude that the use of individual neighborhoods of optimal size provides a generic and thus general approach, since it completely avoids the use of empiric or heuristic knowledge about the scene and/or data which, in turn, would be necessary when specifying neighborhoods in the scope of standard approaches. The use of individual 3D neighborhoods is also in accordance with the idea that the optimal neighborhood size may not be the same for different classes and that it may furthermore depend on the respective point density as well as the local 3D structure. In this regard, the class-specific classification results clearly reveal that, for neighborhood definitions with fixed scale parameter, the suitability may vary from one class to the other (Tables 6.8 and 6.9). In contrast, the proposed method of eigenentropy-based scale selection directly adapts the neighborhood size to the given 3D point cloud data. As a consequence, this method significantly improves the classification results, in particular when considering the mean class recall values. Interestingly, this significant improvement becomes visible for a variety of different classifiers (Tables 6.6 and 6.7). Note that for all these classifiers, the significantly beneficial impact of individual neighborhoods of optimal size on the mean class recall values results from a considerable impact for the smaller classes *wire* and *pole/trunk* (Tables 6.8 and 6.9). Thus, we may state that the respective classifiers are less prone to overfitting when introducing individual neighborhoods of optimal size. This is in accordance with the fact that, for the Oakland 3D Point Cloud Dataset [85, 86], we have an unbalanced test set for which an overall accuracy of 70.5 % could be obtained if only the instances of the class *ground* are correctly classified. Such an extreme overfitting becomes visible when considering the respective mean class recall value of only 20.0 %. Thus, in our experiments, the strongest indicator for the quality of the derived results is represented by the mean class recall value, as only a high overall accuracy may not be sufficient.

For the second component of feature extraction, we have focused on using a feature set which consists of both 3D and 2D features. Whereas the 3D features provide information about the spatial arrangement of neighboring 3D points in terms of linear, planar or volumetric behavior, the projection onto a horizontally oriented plane clearly provides evidence about the presence of building façades, which appear as a line in the 2D projection. If the point density is sufficiently high, we even have an evidence about the presence of curbstone edges. The detection of such man-made structures, in turn, may be an important cue for urban accessibility analysis which is one of the main intentions of current research. Furthermore, the sampling via the discrete 2D accumulation map with quadratic bins introduces a second neighborhood definition with infinite extent in the vertical direction. In this neighborhood definition, the maximal difference and standard deviation of height values provide further insights about the local 3D structure, which are not represented by the other features.

For the third component of feature selection, we have focused on simple and efficient classifier-independent approaches. When considering a Sequential Forward Selection (SFS) according to the feature ranking derived via our proposed general relevance metric, we may observe that the selection of compact and robust subsets of relevant features improves the predictive accuracy of different classifiers (Fig. 6.14) which, in turn, is in accordance with the Hughes phenomenon (see Sect. 6.2.3). Furthermore, it becomes visible that the use of individual neighborhoods of optimal size tends to provide the best classification results for the feature sets derived via different feature selection approaches. A more detailed consideration of the respective results clearly reveals that, in comparison to the complete feature set $\mathscr{S}_{\text{all}}$, the feature set $\mathscr{S}_{\text{dim}}$ consisting of the three dimensionality features L_λ, P_λ, and S_λ is not sufficient in order to obtain adequate classification results (Tables 6.13 and 6.14). This might be due to ambiguities arising from the fact that the classes *wire* and *pole/trunk* provide a linear behavior, while the classes *façade* and *ground* provide a planar behavior. In order to obtain an adequate separation of these classes, additional features have to be taken into account. In this regard, the feature set $\mathscr{S}_{\lambda,\text{3D}}$ of the eigenvalue-based 3D features performs significantly better with respect to both overall accuracy and mean class recall, but the classification results are still not comparable to those obtained for $\mathscr{S}_{\text{all}}$. In contrast, the feature subsets derived via the four filter-based methods for feature selection provide classification results of better quality. Whereas the feature set $\mathscr{S}_{\text{mRMR}}$ performs worst of the filter-based feature selection methods, the feature set $\mathscr{S}_5$ derived via our proposed general relevance metric performs considerably well when taking into account that only 5 features of all 21 features are used (which reduces the required memory for data storage to only about 24 %). The feature sets $\mathscr{S}_{\text{CFS}}$ and $\mathscr{S}_{\text{FCBF}}$ provide a performance close to the performance obtained for the complete feature set $\mathscr{S}_{\text{all}}$ or even better while simultaneously reducing the required memory for data storage to about 29–76 % which, in turn, may represent an important aspect for large-scale considerations. These results are in accordance with the general aim of feature selection to improve the classification results while reducing the computational effort. Since the selection of $\mathscr{S}_{\text{FCBF}}$ is based on heuristics, the feature set $\mathscr{S}_{\text{CFS}}$ derived via Correlation-based Feature Selection provides the method of choice. Note that we only account for filter-based feature selection, since the use of classifier-independent filter-based methods results in simplicity and efficiency compared to other methods interacting with a classifier.

For the fourth component of supervised classification, we may state that the classifiers based on rule-based learning cannot compete with the other classifiers (Tables 6.6, 6.7, 6.8 and 6.9). In contrast, classifiers relying on instance-based learning significantly improve the classification results, but the computational effort for the testing phase is relatively high due to the consideration of a delayed induction process instead of a training phase (Table 6.10). The more sophisticated classifiers based on probabilistic learning, max-margin learning, ensemble learning, and deep learning yield classification results of better quality. However, we have to take into account that max-margin learning, ensemble learning and deep learning require additional time in order to tune the settings of a respective classifier. Thus, the use of Support Vector Machine classifier—for which the computational effort is already high without a parameter tuning (Table 6.10)—does not really satisfy the constraint with respect to efficiency. Since furthermore deep learning relies on heuristically determining the number of nodes in the hidden layer, probabilistic learning and ensemble learning via bagging seem to be favorable. Considering the derived results, the Random Forest classifier provides a good trade-off between accuracy and efficiency. Note that the selection of a Random Forest classifier may further be motivated by its simplicity, since it is relatively easy to understand and use for nonexpert users.

In total, our framework reveals that, based on fully generic solutions, the consideration of optimal neighborhoods improves the classification results in terms of accuracy and less overfitting, whereas the selection of relevant features reduces the computational burden with respect to both processing time and memory consumption without reducing the quality of the classification results. By providing our implementations for neighborhood recovery and feature extraction (in Matlab, C++ and as binaries), we allow end-users to apply the code on their platform and experienced users to involve the code in their investigations. These components may not only be used for point cloud classification, but also for a variety of other applications such as object segmentation or urban accessibility analysis.

6.7.2 Extension Toward Data-Intensive Processing

For the extension of our framework toward data-intensive processing, we may state that, particularly for the smaller classes, a decrease in performance may be observed (Tables 6.16 and 6.17). This might indicate that the respective classes are still not covered representatively for the complexity of urban 3D scenes. However, the derived mean class recall values for both datasets indicate that the recall values across all classes are still relatively high and, accordingly, we may state that the methodology is still less prone to overfitting than standard approaches relying on the use of a fixed scale parameter across all considered 3D points. Consequently, the decrease in performance might mainly arise from the similarity of local 3D structures belonging to respective classes. Additionally, for the Paris-rue-Cassette database, we may observe that vegetation is detected at the balconies (Fig. 6.22) which is in contradiction to the reference labels (Fig. 6.8), but maybe not always in contradiction to the real scene. Note that the Paris-rue-Madame database contains more noise than the Paris-rue-Cassette database (Figs. 6.19 and 6.20) which might be identified as a further reason for relatively low precision values obtained for the respective smaller classes (Table 6.16). In order to increase the precision values, introducing further features and/or multi-scale considerations seems to be necessary. Furthermore, due to the individual classification of 3D points based on only the respective feature vectors, a noisy labeling is to be expected which indeed may be observed in Figs. 6.21 and 6.22.

When focusing on large-scale 3D scene analysis, it is also necessary to consider the computational complexity, where we may for instance consider the processing times required for different subtasks. In the scope of our work, particular interest has to be paid to the tasks of optimal neighborhood size selection and feature extraction, and respective processing times for the separate slices (including the respective padding) are provided in Figs. 6.25 and 6.26. In these figures, the task of optimal neighborhood size selection reveals a linear complexity for increasing numbers of considered 3D points, whereas the task of feature extraction reveals a nonlinear dependency. In the latter case, an in-depth analysis of the implementation indicates that we have a superposition of (i) a linear behavior for calculating 3D features in terms of basic geometric properties or eigenvalue-based features, (ii) a linear behavior for calculating 2D features in terms of basic geometric properties or eigenvalue-based features, and (iii) a nonlinear behavior for calculating 2D features based on the discrete accumulation map.

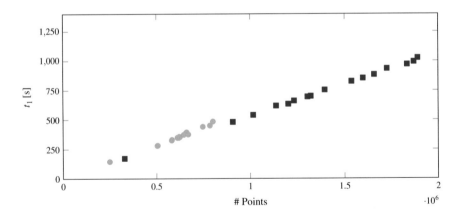

Fig. 6.25 Required processing time per slice for eigenentropy-based scale selection (slices of the Paris-rue-Madame database: *blue squares*; slices of the Paris-rue-Cassette database: *green circles*): a linear behavior may be observed for increasing numbers of 3D points in a slice

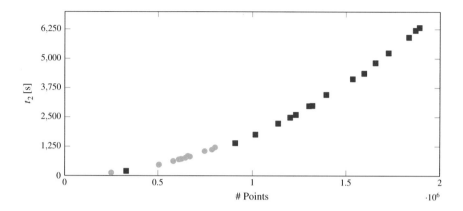

Fig. 6.26 Required processing time per slice for feature extraction (slices of the Paris-rue-Madame database: *blue squares*; slices of the Paris-rue-Cassette database: *green circles*): a nonlinear behavior may be observed for increasing numbers of 3D points in a slice

6.7.3 Extension by Involving Contextual Information

For the extension of our framework toward the use of contextual information, a closer consideration of the overall accuracy of different neighborhood definitions and different classifiers (Table 6.19) reveals that the quality of the classification results is clearly improved when involving contextual information. The lowest overall accuracy is 85.30 % (RF_{100}, $\mathcal{N}_{10}$), whereas the highest overall accuracy is 95.45 % (CRF_{200}^5, $\mathcal{N}_{\text{opt},\lambda}$) and thus more than 10 % better. Furthermore, we may observe that the classification results are better when using the variant RF_{200} of a Random Forest classifier as base classifier delivering the association potentials, and that the use of individual neighborhoods of optimal size for both the extraction of distinctive geometric features and the contextual classification of 3D point cloud data tends to provide the best performance or a performance which is close to the best performance for all considered cases. As a consequence, we may state that the use of individually optimized neighborhoods also improves the results of contextual classification.

Again, we may observe a significantly beneficial impact of individual neighborhoods of optimal size on the mean class recall values (Table 6.20), which results from a considerable impact for the smaller classes *wire* and *pole/trunk* (Tables 6.21 and 6.22). Consequently, we may also state that the involved CRF classifiers are less prone to overfitting when introducing individual neighborhoods of optimal size. Interestingly, the weight w_1 of the interaction potential does not seem to have a significant impact on the classification results, since the effect of changing w_1 in the tested range ($w_1 = 1.0$, $w_1 = 5.0$ and $w_1 = 10.0$) is relatively low compared to the impact of using locally optimized neighborhoods or the impact of using contextual information. Across all derived results, the value of $w_1 = 5.0$ seems to be a good trade-off in this application. The processing time required for contextual classification is in the order of several minutes for training the classifier based on the training set and classifying the test set. For more detailed discussions, we refer to [143].

6.8   Conclusions

In this chapter, we have addressed the issue of 3D scene analysis in terms of uniquely assigning a (semantic) class label to each 3D point of a given 3D point cloud. We have presented a novel, fully automated and versatile framework which involves (i) neighborhood selection, (ii) feature extraction, (iii) feature selection, and (iv) classification. For each component, we have considered a variety of approaches which, in turn, satisfy the constraints of simplicity, efficiency and reproducibility. As main focus of our work, we have considered the interrelated issue of using individual neighborhoods of optimal size in order to extract low-level geometric 3D and 2D features with increased distinctiveness, and we have additionally investigated the impact of selecting a subset consisting of the most relevant features. In a detailed evaluation involving 7 neighborhood definitions, 21 low-level geometric 3D and 2D features, 7 approaches for feature selection, and 10 classifiers, we have demonstrated the significantly beneficial impact of using individual neighborhoods of optimal size as well as the advantages of feature selection in terms of increasing the classification accuracy while simultaneously reducing the computational burden. In particular, the neighborhood selection based on minimizing the measure of eigenentropy over varying scales has provided a significantly positive impact, independent of the respective classifier. Furthermore, those approaches for feature selection which are based on the measure of symmetrical uncertainty for evaluating both feature–class and feature–feature relations have proven to provide the most suitable feature subsets, since they do not only discard irrelevant features but also reduce redundancy among features.

In addition to the main framework, we have presented two further extensions. The first one consists of an extension toward the processing of huge 3D point clouds and takes into account that the selection of individual neighborhoods of optimal size generally increases the computational effort which, in turn, is justified as it significantly improves the classification results in comparison to neighborhoods with a constant scale parameter across all 3D points and furthermore avoids human interaction guided by empiric or heuristic knowledge about the scene and/or data. In this regard, our proposed extension toward data-intensive processing via scene partitioning overcomes the limitation with respect to the computational burden and also allows large-scale 3D scene analysis, which has been demonstrated on two recently published 3D point cloud datasets captured in urban areas. The second extension of our main framework has focused on improving the classification results by introducing a spatially smooth labeling which has been achieved by exploiting contextual information via the use of Conditional Random Fields. Thereby, we have addressed the interrelated issues of (i) neighborhood selection, (ii) feature extraction and (iii) contextual classification, and we have proposed the use of individual 3D neighborhoods of optimal size for the subsequent steps of feature extraction and contextual classification. In this regard, the derived results clearly indicate the beneficial impact of involving contextual information in the classification process and that using individual 3D neighborhoods of optimal size significantly increases the quality of the results for both point-wise and contextual classification. The fact that individually optimized neighborhoods also significantly improve the results for contextual classification mainly arises from the respective impact on the association potentials provided by the base classifier which, in turn, relies on an individual classification of each 3D point by only considering the respective feature vector.

For future work, it would be desirable to carry out deeper investigations concerning the influence of the amount of training data as well as the influence of the number of different classes on the classification results for different datasets. Particularly the number of considered (semantic) classes may have a strong influence on the quality of the classification results, since an increasing number of classes typically causes a higher similarity of these and some classes may even be very hard to distinguish via geometric features. In this regard, the Paris-rue-Cassette database [98] and its extension represent densely sampled urban areas and provide point-wise labels organized in a hierarchy of semantic classes and segmented objects, and these datasets allow objective investigations toward large-scale 3D scene analysis with many different classes. Large geospatial datasets in terms of urban districts, city scale, or even larger are an important prerequisite for the design of large-scale processing workflows which are still in an infant stage with respect to 3D scene analysis as becomes visible in a recent contest on urban 3D point cloud classification [131]. Instead of only focusing on a semantic 3D point cloud classification, a further avenue of research might address a more detailed 3D scene analysis up to object level, e.g., by extracting single objects in a 3D scene such as buildings, trees, cars, or traffic signs, or by exploiting the derived classification results for urban accessibility analysis.

References

1. Aijazi AK, Checchin P, Trassoudaine L (2013) Segmentation based classification of 3D urban point clouds: a super-voxel based approach with evaluation. Remote Sens 5(4):1624–1650
2. Anguelov D, Taskar B, Chatalbashev V, Koller D, Gupta D, Heitz G, Ng A (2005) Discriminative learning of Markov random fields for segmentation of 3D scan data. In: Proceedings of the IEEE computer society conference on computer vision and pattern recognition, vol 2, pp 169–176
3. Arya S, Mount DM, Netanyahu NS, Silverman R, Wu AY (1998) An optimal algorithm for approximate nearest neighbor searching in fixed dimensions. J ACM 45(6):891–923
4. Behley J, Steinhage V, Cremers AB (2012) Performance of histogram descriptors for the classification of 3D laser range data in urban environments. In: Proceedings of the IEEE international conference on robotics and automation, pp 4391–4398
5. Belton D, Lichti DD (2006) Classification and segmentation of terrestrial laser scanner point clouds using local variance information. Int Arch Photogramm Remote Sens Spat Inf Sci XXXVI-5:44–49
6. Blomley R, Weinmann M, Leitloff J, Jutzi B (2014) Shape distribution features for point cloud analysis—A geometric histogram approach on multiple scales. ISPRS Ann Photogramm Remote Sens Spat Inf Sci II-3:9–16

7. Bochow M (2005) Improving class separability—A comparative study of transformation methods for the hyperspectral feature space. In: Proceedings of EARSeL workshop on imaging spectroscopy, pp 439–447
8. Boser BE, Guyon IE, Vapnik VN (1992) A training algorithm for optimal margin classifiers. In: Proceedings of the annual workshop on computational learning theory, pp 144–152
9. Boulch A, Houllier S, Marlet R, Tournaire O (2013) Semantizing complex 3D scenes using constrained attribute grammars. Comput Graph Forum 32(5):33–42
10. Boyko A, Funkhouser T (2011) Extracting roads from dense point clouds in large scale urban environment. ISPRS J Photogramm Remote Sens 66(6):S02–S12
11. Boykov YY, Jolly M-P (2001) Interactive graph cuts for optimal boundary and region segmentation of objects in N-D images. In: Proceedings of the IEEE international conference on computer vision, vol 1, pp 105–112
12. Braun AC (2013) Eine geoökologische und fernerkundliche Prozessanalyse zum Risikozusammenhang zwischen Landnutzung und Biodiversität an einem Beispiel aus Chile. PhD thesis, Institute of Photogrammetry and Remote Sensing, Karlsruhe Institute of Technology (KIT), Karlsruhe, Germany
13. Brédif M, Vallet B, Serna A, Marcotegui B, Paparoditis N (2014) TerraMobilita/IQmulus urban point cloud classification benchmark. In: Proceedings of the IQmulus workshop on processing large geospatial data, pp 1–6
14. Breiman L (1996) Bagging predictors. Mach Learn 4(2):123–140
15. Breiman L (2001) Random forests. Mach Learn 45(1):5–32
16. Bremer M, Wichmann V, Rutzinger M (2013) Eigenvalue and graph-based object extraction from mobile laser scanning point clouds. ISPRS Ann Photogramm Remote Sens Spat Inf Sci II-5/W2:55–60
17. Brodu N, Lague D (2012) 3D terrestrial lidar data classification of complex natural scenes using a multi-scale dimensionality criterion: applications in geomorphology. ISPRS J Photogramm Remote Sens 68:121–134
18. Burges CJC (1998) A tutorial on support vector machines for pattern recognition. Data Min Knowl Discov 2(2):121–167
19. Carlberg M, Gao P, Chen G, Zakhor A (2009) Classifying urban landscape in aerial lidar using 3D shape analysis. In: Proceedings of the IEEE international conference on image processing, pp 1701–1704
20. Ceylan D, Mitra NJ, Zheng Y, Pauly M (2014) Coupled structure-from-motion and 3D symmetry detection for urban facades. ACM Trans Graph 33(1):2:1–15
21. Chang C-C, Lin C-J (2011) LIBSVM: a library for support vector machines. ACM Trans Intell Syst Technol 2(3):27:1–27
22. Chehata N, Guo L, Mallet C (2009) Airborne lidar feature selection for urban classification using random forests. Int Arch Photogramm Remote Sens Spat Inf Sci XXXVIII-3/W8: 207–212
23. Chen C, Liaw A, Breiman L (2004) Using random forest to learn imbalanced data. Technical Report, University of California, Berkeley
24. Cirujeda P, Mateo X, Dicente Y, Binefa X (2014) MCOV: a covariance descriptor for fusion of texture and shape features in 3D point clouds. In: Proceedings of the international conference on 3D vision, pp 551–558
25. Cortes C, Vapnik V (1995) Support-vector networks. Mach Learn 20(3):273–297
26. Cover T, Hart P (1967) Nearest neighbor pattern classification. IEEE Trans Inf Theory 13(1):21–27
27. Criminisi A, Shotton J (2013) Decision forests for computer vision and medical image analysis. Advances in computer vision and pattern recognition, Springer, London
28. Demantké J, Mallet C, David N, Vallet B (2011) Dimensionality based scale selection in 3D lidar point clouds. Int Arch Photogramm Remote Sens Spat Inf Sci XXXVIII-5/W12:97–102
29. Demantké J, Vallet B, Paparoditis N (2012) Streamed vertical rectangle detection in terrestrial laser scans for facade database production. ISPRS Ann Photogramm Remote Sens Spat Inf Sci I-3:99–104

30. Duch W (2006) Filter methods. In: Guyon I, Nikravesh M, Gunn S, Zadeh LA (eds) Feature extraction—Foundations and applications. Studies in fuzziness and soft computing, Springer, Berlin, pp 89–117

31. Efron B (1979) Bootstrap methods: another look at the jackknife. Ann Stat 7(1):1–26

32. Fauvel M (2007) Spectral and spatial methods for the classification of urban remote sensing data. PhD thesis, Grenoble-Image-sPeech-Signal-Automatics Lab, Grenoble Institute of Technology, Grenoble, France

33. Fayyad UM, Irani KB (1993) Multi-interval discretization of continuous-valued attributes for classification learning. In: Proceedings of the international joint conference on artificial intelligence, pp 1022–1027

34. Fehr D, Cherian A, Sivalingam R, Nickolay S, Morellas V, Papanikolopoulos N (2012) Compact covariance descriptors in 3D point clouds for object recognition. In: Proceedings of the IEEE international conference on robotics and automation, pp 1793–1798

35. Filin S, Pfeifer N (2005) Neighborhood systems for airborne laser data. Photogramm Eng Remote Sens 71(6):743–755

36. Fisher RA (1936) The use of multiple measurements in taxonomic problems. Ann Eugen 7(2):179–188

37. Förstner W, Moonen B (2003) A metric for covariance matrices. In: Grafarend EW, Krumm FW, Schwarze VS (eds) Geodesy—The challenge of the 3rd millennium. Springer, Berlin, pp 299–309

38. Freund Y, Schapire RE (1997) A decision-theoretic generalization of on-line learning and an application to boosting. J Comput Syst Sci 55(1):119–139

39. Friedman JH, Bentley JL, Finkel RA (1977) An algorithm for finding best matches in logarithmic expected time. ACM Trans Math Softw 3(3):209–226

40. Frey BJ, MacKay DJC (1998) A revolution: belief propagation in graphs with cycles. In: Proceedings of the neural information processing systems conference, pp 479–485

41. Frome A, Huber D, Kolluri R, Bülow T, Malik J (2004) Recognizing objects in range data using regional point descriptors. In: Proceedings of the European conference on computer vision, vol III, pp 224–237

42. Gerke M, Xiao J (2013) Supervised and unsupervised MRF based 3D scene classification in multiple view airborne oblique images. ISPRS Ann Photogramm Remote Sens Spat Inf Sci II-3/W3:25–30

43. Gini C (1912) Variabilite e mutabilita. Memorie di Metodologia Statistica, pp 25–30

44. Golovinskiy A, Kim VG, Funkhouser T (2009) Shape-based recognition of 3D point clouds in urban environments. In: Proceedings of the IEEE international conference on computer vision, pp 2154–2161

45. Goulette F, Nashashibi F, Abuhadrous I, Ammoun S, Laurgeau C (2006) An integrated on-board laser range sensing system for on-the-way city and road modelling. Int Arch Photogramm Remote Sens Spat Inf Sci XXXVI-1:1–6

46. Gross H, Jutzi B, Thoennessen U (2007) Segmentation of tree regions using data of a full-waveform laser. Int Arch Photogramm Remote Sens Spat Inf Sci XXXVI-3/W49A:57–62

47. Guan H, Li J, Yu Y, Wang C, Chapman M, Yang B (2014) Using mobile laser scanning data for automated extraction of road markings. ISPRS J Photogramm Remote Sens 87:93–107

48. Guo L, Chehata N, Mallet C, Boukir S (2011) Relevance of airborne lidar and multispectral image data for urban scene classification using random forests. ISPRS J Photogramm Remote Sens 66(1):56–66

49. Guo B, Huang X, Zhang F, Sohn G (2014) Classification of airborne laser scanning data using JointBoost. ISPRS J Photogramm Remote Sens 92:124–136

50. Guyon I, Elisseeff A (2003) An introduction to variable and feature selection. J Mach Learn Res 3:1157–1182

51. Guyon I, Gunn S, Nikravesh M, Zadeh LA (2006) Feature extraction: foundations and applications. Springer, Heidelberg

52. Hall MA (1999) Correlation-based feature subset selection for machine learning. PhD thesis, Department of Computer Science, University of Waikato, New Zealand

53. Hebert M, Bagnell JA, Bajracharya M, Daniilidis K, Matthies LH, Mianzo L, Navarro-Serment L, Shi J, Wellfare M (2012) Semantic perception for ground robotics. Proc SPIE 8387(83870Y):1–12
54. Hu H, Munoz D, Bagnell JA, Hebert M (2013) Efficient 3-D scene analysis from streaming data. In: Proceedings of the IEEE international conference on robotics and automation, pp 2297–2304
55. Hughes G (1968) On the mean accuracy of statistical pattern recognizers. IEEE Trans Inf Theory 14(1):55–63
56. John GH, Langley P (1995) Estimating continuous distributions in Bayesian classifiers. In: Proceedings of the conference on uncertainty in artificial intelligence, pp 338–345
57. Johnson AE, Hebert M (1999) Using spin images for efficient object recognition in cluttered 3D scenes. IEEE Trans Pattern Anal Mach Intell 21(5):433–449
58. Jutzi B, Gross H (2009) Nearest neighbour classification on laser point clouds to gain object structures from buildings. Int Arch Photogramm Remote Sens Spat Inf Sci XXXVIII-1-4-7/W5:1–6
59. Khoshelham K, Oude Elberink SJ (2012) Role of dimensionality reduction in segment-based classification of damaged building roofs in airborne laser scanning data. In: Proceedings of the international conference on geographic object based image analysis, pp 372–377
60. Kim HB, Sohn G (2011) Random forests based multiple classifier system for power-line scene classification. Int Arch Photogramm Remote Sens Spat Inf Sci XXXVIII-5/W12:253–258
61. Kim E, Medioni G (2011) Urban scene understanding from aerial and ground lidar data. Mach Vis Appl 22(4):691–703
62. Kolmogorov AN (1933) Grundbegriffe der Wahrscheinlichkeitsrechnung. Springer, Berlin
63. Kononenko I (1994) Estimating attributes: analysis and extensions of RELIEF. In: Proceedings of the European conference on machine learning, pp 171–182
64. Kumar S, Hebert M (2006) Discriminative random fields. Int J Comput Vis 68(2):179–201
65. Lafarge F, Alliez P (2013) Surface reconstruction through point set structuring. Comput Graph Forum 32(2):225–234
66. Lafarge F, Mallet C (2012) Creating large-scale city models from 3D-point clouds: a robust approach with hybrid representation. Int J Comput Vis 99(1):69–85
67. Lafferty JD, McCallum A, Pereira FCN (2001) Conditional random fields: probabilistic models for segmenting and labeling sequence data. In: Proceedings of the international conference on machine learning, pp 282–289
68. Lalonde J-F, Unnikrishnan R, Vandapel N, Hebert M (2005) Scale selection for classification of point-sampled 3D surfaces. In: Proceedings of the international conference on 3-D digital imaging and modeling, pp 285–292
69. Lalonde J-F, Vandapel N, Huber D, Hebert M (2006) Natural terrain classification using three-dimensional ladar data for ground robot mobility. J Field Robot 23(10):839–861
70. Lari Z, Habib A (2012) Alternative methodologies for estimation of local point density index: moving towards adaptive lidar data processing. Int Arch Photogramm Remote Sens Spat Inf Sci XXXIX-B3:127–132
71. Lee I, Schenk T (2002) Perceptual organization of 3D surface points. Int Arch Photogramm Remote Sens Spat Inf Sci XXXIV-3A:193–198
72. Lim EH, Suter D (2009) 3D terrestrial lidar classifications with super-voxels and multi-scale conditional random fields. Comput-Aided Design 41(10):701–710
73. Linsen L, Prautzsch H (2001) Local versus global triangulations. In: Proceedings of Eurographics, pp 257–263
74. Liu H, Motoda H, Setiono R, Zhao Z (2010) Feature selection: an ever evolving frontier in data mining. In: Proceedings of the international workshop on feature selection in data mining, pp 4–13
75. Lodha SK, Kreps EJ, Helmbold DP, Fitzpatrick D (2006) Aerial lidar data classification using support vector machines. In: Proceedings of the international symposium on 3D data processing, visualization, and transmission, pp 567–574

76. Lodha SK, Fitzpatrick DM, Helmbold DP (2007) Aerial lidar data classification using AdaBoost. In: Proceedings of the international conference on 3-D digital imaging and modeling, pp 435–442
77. Lopez de Mantaras R, Armengol E (1998) Machine learning from examples: inductive and lazy methods. Data Knowl Eng 25(1–2):99–123
78. Lu Y, Rasmussen C (2012) Simplified Markov random fields for efficient semantic labeling of 3D point clouds. In: Proceedings of the IEEE/RSJ international conference on intelligent robots and systems, pp 2690–2697
79. Mallet C (2010) Analyse de données lidar à Retour d'Onde Complète pour la classification en milieu urbain. PhD thesis, Télécom Paris Tech, Paris, France
80. Mallet C, Bretar F, Roux M, Soergel U, Heipke C (2011) Relevance assessment of full-waveform lidar data for urban area classification. ISPRS J Photogramm Remote Sens 66(6):S71–S84
81. Melgani F, Bruzzone L (2004) Classification of hyperspectral remote sensing images with support vector machines. IEEE Trans Geosci Remote Sens 42(8):1778–1790
82. Mitra NJ, Nguyen A (2003) Estimating surface normals in noisy point cloud data. In: Proceedings of the annual symposium on computational geometry, pp 322–328
83. Monnier F, Vallet B, Soheilian B (2012) Trees detection from laser point clouds acquired in dense urban areas by a mobile mapping system. ISPRS Ann Photogramm Remote Sens Spat Inf Sci I-3:245–250
84. Muja M, Lowe DG (2009) Fast approximate nearest neighbors with automatic algorithm configuration. In: Proceedings of the international conference on computer vision theory and applications, vol 1, pp 331–340
85. Munoz D, Vandapel N, Hebert M (2008) Directional associative Markov network for 3-D point cloud classification. In: Proceedings of the international symposium on 3D data processing, visualization and transmission, pp 63–70
86. Munoz D, Bagnell JA, Vandapel N, Hebert M (2009) Contextual classification with functional max-margin Markov networks. In: Proceedings of the IEEE computer society conference on computer vision and pattern recognition, pp 975–982
87. Munoz D, Vandapel N, Hebert M (2009) Onboard contextual classification of 3-D point clouds with learned high-order Markov random fields. In: Proceedings of the IEEE international conference on robotics and automation, pp 2009–2016
88. Musialski P, Wonka P, Aliaga DG, Wimmer M, Van Gool L, Purgathofer W (2013) A survey of urban reconstruction. Comput Graph Forum 32(6):146–177
89. Najafi M, Taghavi Namin S, Salzmann M, Petersson L (2014) Non-associative higher-order Markov networks for point cloud classification. In: Proceedings of the European conference on computer vision, vol V, pp 500–515
90. Niemeyer J, Wegner JD, Mallet C, Rottensteiner F, Soergel U (2011) Conditional random fields for urban scene classification with full waveform lidar data. In: Stilla U, Rottensteiner F, Mayer H, Jutzi B, Butenuth M (eds) Photogrammetric image analysis, ISPRS Conference—Proceedings. Lecture notes in computer science, vol 6952, Springer, Heidelberg, pp 233–244
91. Niemeyer J, Rottensteiner F, Soergel U (2012) Conditional random fields for lidar point cloud classification in complex urban areas. ISPRS Ann Photogramm Remote Sens Spat Inf Sci I-3:263–268
92. Niemeyer J, Rottensteiner F, Soergel U (2014) Contextual classification of lidar data and building object detection in urban areas. ISPRS J Photogramm Remote Sens 87:152–165
93. Nurunnabi A, Belton D, West G (2012) Robust segmentation in laser scanning 3D point cloud data. Proceedings of the international conference on digital image computing techniques and applications, pp 1–8
94. Osada R, Funkhouser T, Chazelle B, Dobkin D (2002) Shape distributions. ACM Trans Graph 21(4):807–832
95. Oude Elberink S, Kemboi B (2014) User-assisted object detection by segment based similarity measures in mobile laser scanner data. Int Arch Photogramm Remote Sens Spat Inf Sci XL-3:239–246

96. Özuysal M, Fua P, Lepetit V (2007) Fast keypoint recognition in ten lines of code. In: Proceedings of the IEEE computer society conference on computer vision and pattern recognition, pp 1–8

97. Paparoditis N, Papelard J-P, Cannelle B, Devaux A, Soheilian B, David N, Houzay E (2012) Stereopolis II: a multi-purpose and multi-sensor 3D mobile mapping system for street visualisation and 3D metrology. Revue Française de Photogrammétrie et de Télédétection 200:69–79

98. Paparoditis N, Vallet B, Marcotegui B, Serna A (2014) IQmulus & TerraMobilita contest—Analysis of mobile laser scans (MLS) in dense urban environments. MATIS laboratory, French national mapping agency (IGN) and Center for Mathematical Morphology (CMM), MINES ParisTech, Paris. http://data.ign.fr/benchmarks/UrbanAnalysis/. Accessed 30 May 2015

99. Pauly M, Keiser R, Gross M (2003) Multi-scale feature extraction on point-sampled surfaces. Comput Graph Forum 22(3):281–289

100. Pearson K (1896) Mathematical contributions to the theory of evolution. III. Regression, heredity and panmixia. Philos Trans Royal Soc Lond A(187):253–318

101. Peng H, Long F, Ding C (2005) Feature selection based on mutual information criteria of max-dependency, max-relevance, and min-redundancy. IEEE Trans Pattern Anal Mach Intell 27(8):1226–1238

102. Poullis C, You S (2009) Automatic reconstruction of cities from remote sensor data. In: Proceedings of the IEEE computer society conference on computer vision and pattern recognition, pp 2775–2782

103. Press WH, Flannery BP, Teukolsky SA, Vetterling WT (1988) Numerical recipes in C: the art of scientific computing. Cambridge University Press, Cambridge

104. Pu S, Rutzinger M, Vosselman G, Oude Elberink S (2011) Recognizing basic structures from mobile laser scanning data for road inventory studies. ISPRS J Photogramm Remote Sens 66(6):S28–S39

105. Quinlan JR (1986) Induction of decision trees. Mach Learn 1(1):81–106

106. Riedmiller M, Braun H (1993) A direct adaptive method for faster backpropagation learning: the RPROP algorithm. In: Proceedings of the IEEE international conference on neural networks, pp 586–591

107. Rumelhart DE, Hinton GE, Williams RJ (1986) Learning representations by back-propagating errors. Nature 323:533–536

108. Rusu RB (2009) Semantic 3D object maps for everyday manipulation in human living environments. PhD thesis, Computer Science department, Technische Universität München

109. Rusu RB, Marton ZC, Blodow N, Beetz M (2008) Persistent point feature histograms for 3D point clouds. In: Proceedings of the international conference on intelligent autonomous systems, pp 119–128

110. Rusu RB, Blodow N, Beetz M (2009) Fast point feature histograms (FPFH) for 3D registration. In: Proceedings of the IEEE international conference on robotics and automation, pp 3212–3217

111. Saeys Y, Inza I, Larrañaga P (2007) A review of feature selection techniques in bioinformatics. Bioinform 23(19):2507–2517

112. Samet H (2006) Foundations of multidimensional and metric data structures. Morgan Kaufmann, St. Louis

113. Schapire RE (1990) The strength of weak learnability. Mach Learn 5(2):197–227

114. Schindler K (2012) An overview and comparison of smooth labeling methods for land-cover classification. IEEE Trans Geosci Remote Sens 50(11):4534–4545

115. Schmidt A, Rottensteiner F, Soergel U (2012) Classification of airborne laser scanning data in Wadden Sea areas using conditional random fields. Int Arch Photogramm Remote Sens Spat Inf Sci XXXIX-B3:161–166

116. Schmidt A, Rottensteiner F, Soergel U (2013) Monitoring concepts for coastal areas using lidar data. Int Arch Photogramm Remote Sens Spat Inf Sci XL-1/W1:311–316

117. Schmidt A, Niemeyer J, Rottensteiner F, Soergel U (2014) Contextual classification of full waveform lidar data in the Wadden Sea. IEEE Geosci Remote Sens Lett 11(9):1614–1618

118. Secord J, Zakhor A (2007) Tree detection in urban regions using aerial lidar and image data. IEEE Geosci Remote Sens Lett 4(2):196–200
119. Serna A, Marcotegui B (2013) Urban accessibility diagnosis from mobile laser scanning data. ISPRS J Photogramm Remote Sens 84:23–32
120. Serna A, Marcotegui B (2014) Detection, segmentation and classification of 3D urban objects using mathematical morphology and supervised learning. ISPRS J Photogramm Remote Sens 93:243–255
121. Serna A, Marcotegui B, Goulette F, Deschaud J-E (2014) Paris-rue-Madame database: a 3D mobile laser scanner dataset for benchmarking urban detection, segmentation and classification methods. In: Proceedings of the international conference on pattern recognition applications and methods, pp 819–824
122. Shannon CE (1948) A mathematical theory of communication. Bell Syst Tech J 27(3): 379–423
123. Shapovalov R, Velizhev A, Barinova O (2010) Non-associative Markov networks for 3D point cloud classification. Int Arch Photogramm Remote Sens Spat Inf Sci XXXVIII-3A:103–108
124. Shapovalov R, Velizhev A (2011) Cutting-plane training of non-associative Markov network for 3D point cloud segmentation. In: Proceedings of the IEEE international conference on 3D digital imaging, modeling, processing, visualization and transmission, pp 1–8
125. Shapovalov R, Vetrov D, Kohli P (2013) Spatial inference machines. In: Proceedings of the IEEE computer society conference on computer vision and pattern recognition, pp 2985–2992
126. Shotton J, Winn J, Rother C, Criminisi A (2009) TextonBoost for image understanding: multi-class object recognition and segmentation by jointly modeling texture, layout, and context. Int J Comput Vis 81(1):2–23
127. Tokarczyk P, Wegner JD, Walk S, Schindler K (2013) Beyond hand-crafted features in remote sensing. ISPRS Ann Photogramm Remote Sens Spat Inf Sci II-3/W1:35–40
128. Tombari F, Salti S, Di Stefano L (2010) Unique signatures of histograms for local surface description. In: Proceedings of the European conference on computer vision, vol III, pp 356–369
129. Tombari F, Fioraio N, Cavallari T, Salti S, Petrelli A, Di Stefano L (2014) Automatic detection of pole-like structures in 3D urban environments. In: Proceedings of the IEEE/RSJ international conference on intelligent robots and systems, pp 4922–4929
130. Unnikrishnan R, Hebert M (2008) Multi-scale interest regions from unorganized point clouds. In: Proceedings of the IEEE computer society conference on computer vision and pattern recognition workshops, pp 1–8
131. Vallet B, Brédif M, Serna A, Marcotegui B, Paparoditis N (2015) TerraMobilita/IQmulus urban point cloud analysis benchmark. Comput Graph 49:126–133
132. Vanegas CA, Aliaga DG, Benes B (2012) Automatic extraction of Manhattan-world building masses from 3D laser range scans. IEEE Trans Vis Comput Graph 18(10):1627–1637
133. Velizhev A, Shapovalov R, Schindler K (2012) Implicit shape models for object detection in 3D point clouds. ISPRS Ann Photogramm Remote Sens Spat Inf Sci I-3:179–184
134. Vosselman G (2013) Point cloud segmentation for urban scene classification. Int Arch Photogramm Remote Sens Spat Inf Sci XL-7/W2:257–262
135. Vosselman G, Klein R (2010) Visualisation and structuring of point clouds. In: Vosselman G, Maas H-G (eds) Airborne and terrestrial laser scanning. Whittles Publishing, Dunbeath, pp 45–81
136. Waldhauser C, Hochreiter R, Otepka J, Pfeifer N, Ghuffar S, Korzeniowska K, Wagner G (2014) Automated classification of airborne laser scanning point clouds. In: Koziel S, Leifsson L, Yang X-S (eds) Solving computationally expensive engineering problems—Methods and applications. Springer, New York, pp 269–292
137. Wegner JD, Soergel U, Rosenhahn B (2011) Segment-based building detection with conditional random fields. In: Proceedings of the joint urban remote sensing event, pp 205–208
138. Weinmann M, Jutzi B, Mallet C (2013) Feature relevance assessment for the semantic interpretation of 3D point cloud data. ISPRS Ann Photogramm Remote Sens Spat Inf Sci II-5/W2:313–318

139. Weinmann M, Jutzi B, Mallet C (2014): Semantic 3D scene interpretation: a framework com-
    bining optimal neighborhood size selection with relevant features. ISPRS Ann Photogramm
    Remote Sens Spat Inf Sci II-3:181–188
140. Weinmann M, Jutzi B, Mallet C (2014) Describing Paris: automated 3D scene analysis via dis-
    tinctive low-level geometric features. In: Proceedings of the IQmulus workshop on processing
    large geospatial data, pp 1–8
141. Weinmann M, Urban S, Hinz S, Jutzi B, Mallet C (2015) Distinctive 2D and 3D features for
    automated large-scale scene analysis in urban areas. Comput Graph 49:47–57
142. Weinmann M, Jutzi B, Hinz S, Mallet C (2015) Semantic point cloud interpretation based
    on optimal neighborhoods, relevant features and efficient classifiers. ISPRS J Photogramm
    Remote Sens 105:286–304
143. Weinmann M, Schmidt A, Mallet C, Hinz S, Rottensteiner F, Jutzi B (2015) Contextual
    classification of point cloud data by exploiting individual 3D neighborhoods. ISPRS Ann
    Photogramm Remote Sens Spat Inf Sci II-3/W4:271–278
144. Weinmann M, Mallet C, Hinz S, Jutzi B (2015) Efficient interpretation of 3D point clouds by
    assessing feature relevance. AVN–Allg Vermess-Nachr 10(2015):308–315
145. West KF, Webb BN, Lersch JR, Pothier S, Triscari JM, Iverson AE (2004) Context-driven
    automated target detection in 3-D data. Proc SPIE 5426:133–143
146. Wurm KM, Kretzschmar H, Kümmerle R, Stachniss C, Burgard W (2014) Identifying vege-
    tation from laser data in structured outdoor environments. Robot Auton Syst 62(5):675–684
147. Xiong X, Munoz D, Bagnell JA, Hebert M (2011) 3-D scene analysis via sequenced predictions
    over points and regions. In: Proceedings of the IEEE international conference on robotics and
    automation, pp 2609–2616
148. Xu S, Oude Elberink SJ, Vosselman G (2012) Entities and features for classification of airborne
    laser scanning data in urban area. ISPRS Ann Photogramm Remote Sens Spat Inf Sci I-4:
    257–262
149. Xu S, Vosselman G, Oude Elberink S (2014) Multiple-entity based classification of airborne
    laser scanning data in urban areas. ISPRS J Photogramm Remote Sens 88:1–15
150. Yokoyama H, Date H, Kanai S, Takeda H (2013) Detection and classification of pole-like
    objects from mobile laser scanning data of urban environments. Int J CAD/CAM 13(2):
    31–40
151. Yu L, Liu H (2003) Feature selection for high-dimensional data: a fast correlation-based filter
    solution. In: Proceedings of the international conference on machine learning, pp 856–863
152. Yu T-H, Woodford OJ, Cipolla R (2013) A performance evaluation of volumetric 3D interest
    point detectors. Int J Comput Vis 102(1–3):180–197
153. Zhao Z, Morstatter F, Sharma S, Alelyani S, Anand A, Liu H (2010) Advancing feature
    selection research—ASU feature selection repository. Technical report, School of Computing,
    Informatics, and Decision Systems Engineering, Arizona State University, Tempe
154. Zheng Q, Sharf A, Wan G, Li Y, Mitra NJ, Cohen-Or D, Chen B (2010) Non-local scan
    consolidation for 3D urban scenes. ACM Trans Graph 29(4):94:1–9
155. Zhou Q-Y, Neumann U (2009) A streaming framework for seamless building reconstruction
    from large-scale aerial lidar data. In: Proceedings of the IEEE computer society conference
    on computer vision and pattern recognition, pp 2759–2766
156. Zhou Q-Y, Neumann U (2013) Complete residential urban area reconstruction from dense
    aerial lidar point clouds. Graph Model 75(3):118–125
157. Zhou L, Vosselman G (2012) Mapping curbstones in airborne and mobile laser scanning data.
    Int J Appl Earth Obs Geoinf 18(1):293–304

Chapter 7
Conclusions and Future Work

In this book, we have addressed the fully automatic processing and analysis of 3D point clouds, and we have presented novel concepts and methods for the digitization, reconstruction, interpretation, and understanding of static and dynamic indoor and outdoor scenes. These concepts and methods have been integrated in a framework for advanced 3D point cloud processing from raw 3D point cloud data to semantic objects in the scene. As main components of the presented framework, we have addressed (i) the filtering of noisy data, (ii) the extraction of appropriate features, (iii) the adequate alignment of several 3D point clouds in a common coordinate frame, (iv) the enrichment of 3D point cloud data with other types of information, and (v) the semantic interpretation of 3D point clouds. Based on an extensive evaluation of different components of our framework on various benchmark datasets, we have provided results which clearly demonstrate the performance of the proposed approaches in comparison to state-of-the-art techniques and their great potential for a variety of applications.

Besides reflecting fundamentals and giving a comprehensive survey on related work, we have presented a variety of scientific contributions amongst which the following ones represent the key contributions presented in this book:

- **A geometric quantification of the quality of range measurements**:
  In order to assess the quality of scanned 3D points, we have presented (i) the measure of *range reliability* and (ii) the measure of *local planarity*. Both of them allow an appropriate filtering of raw 3D point cloud data by evaluating local characteristics in the respective 2D image representation in the form of a range image. While a filtering based on the measure of range reliability relies on a user-defined threshold and may thus be more or less strict, a filtering based on the measure of local planarity provides a generic solution which generally tends to be rather strict. However, both of these measures offer an adequate handling of unreliable range measurements corresponding to object edges, which represents an important issue not taken into account when only using intensity information for filtering raw 3D point cloud data. Besides their suitability for point quality assessment, the novel measures have been involved in the context of point cloud registration

© Springer International Publishing Switzerland 2016
M. Weinmann, *Reconstruction and Analysis of 3D Scenes*,
DOI 10.1007/978-3-319-29246-5_7

in order to define a consistency check which allows to filter unreliable feature correspondences between the image representations of different scans and thus to facilitate an efficient and robust registration of 3D point clouds by retaining only those feature correspondences with reliable range information.

- **A shift of paradigms in point cloud registration**:
  In order to appropriately align several 3D point clouds in a common coordinate frame, we have focused on extracting local features in the respective intensity images and a subsequent forward projection to 3D space by exploiting the corresponding range information. Since local features are characterized by keypoints representing the respective image locations, the respective approaches for point cloud registration may be referred to as *keypoint-based point cloud registration approaches*. While respective approaches have been involved in the literature for years, only little attention has been paid to the suitability of the selected features or the registration approach itself. Regarding the selected features, respective investigations mainly focused on the use of specific types of local features representing the most prominent approaches for deriving feature correspondences. Instead, our investigations have involved a *variety of feature detector–descriptor combinations* and thereby revealed that more recent approaches for extracting local features significantly alleviate point cloud registration in terms of robustness, efficiency, and accuracy. Regarding the selected registration approach, recent investigations typically rely on solving the task of keypoint-based point cloud registration in the classic way by estimating the rigid Euclidean transformation between sets of corresponding 3D points. Instead, we have taken into account that a rigid transformation only addresses 3D cues in terms of how good corresponding 3D points fit together with respect to the Euclidean distance, whereas feature correspondences between 2D image representations in the form of range and intensity images also allow to involve 2D cues in terms of how good the 2D locations in the image representations fit together. As a result of taking into account both 3D and 2D cues, an improvement in accuracy may be expected for the registration results. In order to take into account both 3D and 2D cues, we have hence proposed a shift of paradigms in point cloud registration by transferring the task of point cloud registration to (i) the task of solving the Perspective-n-Point (PnP) problem, where we have presented a *projective scan matching* which relies on the use of 3D/2D correspondences in order to solve the PnP problem and (ii) the task of solving the relative orientation problem, where we have presented an *omnidirectional scan matching* which relies on the relative pose estimation based on sets of bearing vectors. Thereby, we have also introduced our proposed point quality measures in order to increase the robustness of point cloud registration.

- **An enrichment of existing 3D point cloud data by thermal information**:
  In order to enrich existing 3D point cloud data by thermal information, we have presented (i) an approach relying on a *homography-based image registration* in order to appropriately co-register 3D point cloud data and thermal infrared images for almost planar scenes and (ii) an approach relying on a *projective scan matching* in order to accurately co-register 3D point cloud data and thermal infrared

images for general scenes without regular surface structures. Since the approach relying on a homography-based image registration makes strong assumptions on the scene structure which are for instance satisfied when considering walls or building façades, inaccuracies in the data co-registration may be expected if these assumptions are not satisfied. In contrast, the approach relying on a projective scan matching may be applied without making strong assumptions on the 3D scene structure and therefore outperforms the approach relying on a homography-based image registration in terms of both accuracy and applicability.

- **A framework for the semantic interpretation of 3D point cloud data**:
  In order to conduct a 3D scene analysis in terms of uniquely assigning each 3D point of the considered 3D point cloud a respective (semantic) class label, we have presented a framework for the semantic interpretation of 3D point cloud data. This framework is composed of four components addressing (i) neighborhood selection, (ii) feature extraction, (iii) feature selection, and (iv) classification. While recent investigations on 3D scene analysis have mainly addressed the classification scheme, we have extensively investigated further potential sources for improvements with respect to the classification accuracy and our contributions in this regard are manifold. We have presented (i) the strategy of *eigenentropy-based scale selection* in order to assign a local neighborhood of optimal size to each individual 3D point of a given 3D point cloud and thus obtain geometric features with increased distinctiveness, (ii) the idea of a *feature relevance assessment* for selecting only the most relevant features among a variety of low-level geometric 3D and 2D features in order to improve both predictive accuracy and computational efficiency of 3D scene analysis, (iii) an extension of the main framework toward data-intensive processing in order to address *large-scale 3D scene analysis* and (iv) an extension of the main framework toward the *use of contextual information* in order to account for probably correlated labels among neighboring 3D points and thus obtain a further improvement of the classification results.

The methodology developed in the scope of our work addresses different research topics as well as respective challenges of recent research. Accordingly, future work may address a variety of issues reaching from a benchmarking of methods on publicly available datasets via improvements with respect to the single components of the framework to the consideration of interactions within the framework.

- **Benchmarking**:
  For each component of the presented framework for advanced 3D point cloud processing, it would be desirable to integrate further approaches and to perform an extensive evaluation on a variety of publicly available datasets with respective ground truth annotations. Thereby, different criteria might also be specified by potential end-users in order to allow them to select appropriate methods according to their requirements.
- **Extended point cloud registration**:
  While we have focused on an automated organization of TLS scans and a successive pair-wise registration, it would be desirable to conduct a subsequent global registration or multiview registration in terms of globally optimizing the alignment

of all respective 3D point clouds by simultaneously minimizing registration errors with respect to all available scans (e.g., via bundle adjustment). Moreover, the presented approaches for point cloud registration may be used in order to estimate the position of a moving sensor platform equipped with range cameras. In this regard, it would be possible to further increase the robustness of the presented approaches by for instance involving state estimation methods such as an extended Kalman filter (EKF) or a particle filter. Besides such conceptual extensions, it would also be possible to consider more complex environments (e.g., natural environments such as forests, agricultural areas, or areas along a river channel, but also narrow environments such as mine tunnels or tunnels constructed for road or rail traffic) and 3D point clouds comprising a significantly larger number of 3D points.

- **Extended data co-registration**:
  While we have focused on a co-registration of 3D point cloud data and thermal 2D imagery for a static sensor platform and a static or dynamic scene, it would be interesting to address an extension toward moving sensor platforms and both static and dynamic scenes. In this context, it would also be desirable to integrate further sensor types such as digital cameras, multispectral cameras, etc. in the acquisition system in order to gain further complementary information which could facilitate scene analysis.

- **Extended point-based scene interpretation**:
  While we have focused on 3D scene analysis in terms of assigning each 3D point of a given 3D point cloud a respective (semantic) class label and thereby considering large 3D point clouds but only a relatively small number of less than 10 classes, it would be desirable to address large-scale 3D scene analysis with many different classes which is still in an infant stage. Furthermore, it would be possible to conduct a more detailed scene analysis up to object level by for instance extracting buildings, trees, cars, or traffic signs which, in turn, could be used for subsequent tasks such as deriving 3D models with different levels of detail or carrying out an urban accessibility analysis in a fully automatic manner. In this regard, the step from a classified 3D point cloud toward single objects in the scene could involve smooth labeling techniques or context models, but also a segmentation based on the results of point cloud classification.

- **Segment-based scene interpretation**:
  While we have focused on a point-based scene interpretation in terms of assigning each 3D point of a given 3D point cloud a respective (semantic) class label, it would also be possible to directly conduct a segment-based scene interpretation by (i) deriving segments from the 3D point cloud data (e.g., via k-means algorithm, mean shift algorithm, or graph cuts), (ii) describing characteristic properties of the derived segments (e.g., via those features based on the 3D structure tensor), and (iii) classifying the derived segments based on the segment-wise feature vectors (e.g., via a Support Vector Machine or a Random Forest classifier) in order to conduct a more detailed scene analysis up to object level.

- **Interactions within the presented framework**:
  While we have focused on a separate consideration of each component of the presented framework for advanced 3D point cloud processing, it would also be possible to extend the framework by considering interactions between the components for (i) point cloud registration, (ii) co-registration of 3D and 2D data, and (iii) 3D scene analysis. In this regard, it would for instance be possible to exploit results of 3D scene analysis for improving the results of point cloud registration and data co-registration, while the results of data co-registration could be used in order to obtain further point-wise features and thus possibly improved results for 3D scene analysis.

• Interactions within the presented framework

Index

© Springer International Publishing Switzerland 2016
M. Weinmann, *Reconstruction and Analysis of 3D Scenes*,
DOI 10.1007/978-3-319-29246-5